METAL TECHNOLOGY IN MEDIEVAL INDIA

METAL TECHNOLOGY
IN
MEDIEVAL INDIA

SAYED JAFAR MAHMUD

1988

DAYA PUBLISHING HOUSE
DELHI-110006.

ISBN 81-7035-039-5

Published by: Mr Anil Mittal for Daya Publishing House
at 1302, Vaid Wada, Nai Sarak, Delhi-110006, Phone : 260116.
Printed by: Arihanta Printers, Naveen, Shahadra, Delhi-32.
Bound at Bookwrap, Delhi.

PRINTED IN INDIA

Foreword

This short but useful tract by Jafar Mahmud is a welcome addition to create an awareness and understanding of India's tradition of metal technology—its development as well as its spread in different parts of the country. The usefulness of the tract is enhanced by the detailed discussion of the different mines, in different parts of the country, the method of extraction, the composition of different metals in their mined forms, the tools used and the methods of refining. Further, it also describes the manufacture of different objects.

Two features of Jafar Mahmud's tract are worth mentioning. Firstly, he gives a brief description of mines, mining and products from historical records. In doing so, he also provides scientific explanation, wherever it is needed in the constitution of mined material as well as the changes it undergoes in refining and making products.

Secondly, having visited the sites and the working of the processes personally, he provides an authentic account, which is not merely extracted from historical and other records. This is particularly so in the case of tribal tradition, which he has looked into in some detail.

I hope this tract would lead to more intensive study of the India's tradition in the field of metal technology, both in the historical context of and compared to that of other countries and as well as in its scientific context and technological practices and their evolution.

A. Rahman

Preface

There is a world-wide interest in the study of history of science and technology as one important aspect in understanding man's cultural patterns. The history of metals is a specialized one and tends to be shut away from the general reader in rather grim-looking manuscripts and papers. This is a pity, because metals retain the attraction and wonder they had for us in our youth and they are of great importance to mankind.

There has been, as yet no single comprehensive compilation devoted to the saga of Metal Technology in Medieval India. Therefore, an earnest attempt has been made in this monograph to meet the prima facie need and to present a total picture of the metal Technology in India during the Medieval period.

In preparation of this monograph, recourse has, naturally, been taken to the contemporary and near contemporary sources available both in India and abroad. Further, information has been collected through metallurgical analysis of various samples. While every effort has been made to ensure proper presentation, many unintended shortcomings, if detected and pointed out by readers, may be removed from the next edition.

S. Jafar Mahmud

Contents

Introduction

Although metal-working was not the first craft known to mankind, our present material civilization is derived from our knowledge and use of metals. Metals and metal-working have conditioned the development of human culture throughout the ages. Generally, utilization of metals affects the society in two ways. From one standpoint the technology of metal helped the change in the basic techniques of economy and livelihood. In this respect mention may be made of its influence on agriculture and methods of warfare. Control over the use of harder type of metals empowered ancient societies to overpower their neighbouring communities. And we know that conquest of all kinds help the spread of ideas. From the realm of ideas we could easily move to other aspects of metals in early civilization. It may be remembered that change of materials does not confine itself to a change in technology. Introduction of new materials provides the artists and craftsmen with new opportunities to enlarge the scope of their art. The same thing happened in case of metals. Metals provided and opened up new avenues of artistic ideas in a given society.

From the time of Tubal Cain onwards, the early users of metals would probably have described them as bright substances which were hard but could be hammered into various useful shapes. All the metals which they knew were heavy—iron, for instance, being nearly three times as heavy as granite. If the metal-worker was also a hunter of warrior he might speak with satisfaction of the sharpness of metal weapons and of their endurance through many combats. Craftsmen in wood or in stone would praise the new metal cutting-tools which made their jobs so much easier. Thus, even in the very early days, there grew an appreciation of the value of metals.

Man learned to make fires hot enough to melt metals in earthenware containers, which the Romans called "crucibuli" and which we now call "crucibles". He discovered that the molten metal could be poured into the cavity made by placing together two halves of a hollowed-out clay or stone mould; the metal filled the cavity and, when solid, it was found to have taken the shape of the cavity. Archaeologists have discovered ancient bronze swords and arrow-heads, made by casting in this manner.

The art of blending metals was gradually developed and it became known that an 'alloy' formed in this way was sometimes stronger, harder, and tougher than the metals of which it was composed. Probably, the first alloy to be made was a bronze, consisting of copper, with about one part in ten of tin. The primitive metallurgists discovered that if a greater proportion of tin were used the alloy was harder, while less tin gave a softer alloy, so that for different purposes bronzes with varying tin contents were deliberately produced. By the time the Romans came to Britain, they were using iron and bronze for weapons, tools, and farming implements; copper for vessels and ornaments; lead for water pipes, baths and even coffins; tin, gold, and silver for ornaments; and silver, brass, and bronze for coinage.

Gold and, to a lesser extent, silver, were called 'noble metals' because they could be exposed to the atmosphere for a long time without tarnishing and because they could be melted repeatedly without much loss in weight. These characteristics led to their being used for jewellery and eventually coinage. The possession of noble metals consequently became a measure of wealth so that gold and silver were coveted for their monetary as distinct from their utilitarian value. All the other metals then known, such as tin, lead, copper and iron, were by contrast considered 'base metals'.

Owing to the general similarities between all metals it was perhaps natural to imagine that one metal could be changed into another, and it seemed particularly desirable that base metals should be transmuted into noble ones—gold for preference. Beginning in the Middle East in the early part of the Christian era, and thriving in that part of the world and in Europe till the end of the seventeenth century, the art of alchemy became associated with the search for the elusive Philosophers'

Stone, which was supposed to be capable of turning base metals into gold or silver.

As recently as the end of the eighteenth century a writer defined alchemy as 'a science and art of making a fermentative powder which transmutes imperfect metals into gold and which serves as a universal remedy for all the ills of man, animals, and plants'. The Philosopher Francis Bacon was nearer to the truth when he commented; 'Alchemy may be compared to the man who told his sons that he had left them gold buried somewhere in his vineyard; where they by digging found no gold, but by turning the mould about the roots of the vines, procured a plentiful vintage. So the search and endeavours to make gold have brought many useful inventions and instructive experiments to light'.

While the alchemists and their wealthy, and often exasperated, patrons were trying all kinds of experiments to produce the Philosophers' Stone, the metal-workers, though using fewer incantations, were developing almost as wonderful a process— the changing of dull, earthy minerals or 'ores' into metals by smelting them with charcoal in a fire or furnace. They learned how to recognize those metallic ores which could be smelted profitably and how to transform them into metal. It is not surprising that their efforts were sometimes unsuccessful. Even today, producers of metal encounter difficulties because of the presence of impurities in the ore, which upset smelting operations or have a hurmful influence on the resulting metal. In those early days such happenings were freely attributed to the Evil Eye, or to the attention of hobgoblins.

1

History of Iron and Steel

Iron in Antiquity

Iron finds mention in ancient Indian texts. It has been universally accepted that the Rg Veda is the earliest collection of hymns in Sanskrit. In the Rg Veda, the word Ayas has been mentioned several times. It would appear that this word meant one of the metals, namely, iron, copper or gold in different contexts. It has, however, been pointed out that Ayas would only mean iron and not copper or gold, and also Loha would refer to copper and Shyamam to iron. The latter word has been used in the Atharva Veda also. In order to differentiate between copper and iron, the latter metal has been referred to by scholars as Krishnayas. In the later Vedic Age, the word Ayas was used with different adjectives prefixed to mean either copper or iron. Thus, Lohayas or Lohitayas meant copper, whereas Shyamayas or Karsnayas signified iron. Gold was looked upon as a mark of nobility and iron as a mark of the common people. In the light of the reference to iron in the Vedas, it would be evident that iron was known in the early Vedic times in India and came to be identified clearly as time progressed.

In order to gain an idea of the time frame with regard to the early use of iron, it is necessary to know the age of the Vedas, which has been fixed by scholars. It has been stated that the upper limit for the period of the Vedas is the date of the oldest hymns of the Rg Veda. McDonell and Keith are of the opinion that this date is not much earlier than 1200 B.C. It is generally believed that Asia Minor or the Caucasus was the origin, where probably the first smelting operations of iron were carried out and the Hittites knew the smelting of iron between 1800 and 1200 B.C. By about 1000 B.C. there was extensive use of iron in the near East.

In an exhaustive study, a clue to the discovery of iron making has been pointed out. Amongst the Arundel Marbles now at Oxford, England, is a tablet known as the Parian Chronicle inscribed in 260 B.C. One revealing line reads. "From the time when Minos,.... built Apollonia and iron was discovered in Ida, by the Dactyls and Kelmis during the reign of Pandaion of Athens...". Since it is established that Pandaion, the Athenian King, ruled about 1450 B.C..., this can be taken as one approximate date for man's first use of iron.

However, there were other capable metal workers, such as the Chalybes in various regions of Asia Minor. If the discovery of iron making could be dated around 1450 B.C. in Asia Minor, and if Rg Veda could not be placed earlier than 1200 B.C. according to European scholars, then it could be that the race which colonised India at a very early age, could have been in possession of the art of working iron. This view finds admirable corroboration from the fact that, according to a recent study based on literary and archaeological evidence, c 1200 B.C. could be taken as the critical period for the utilitarian usages of iron in the Eastern Mediterranean, though iron was known to have been used sparingly earlier for votive and amuletic purposes.

Contrary to common belief that copper was the first metal extracted by man from ore, it has been stated that lead was the first metal to be extracted about 9000 years ago. A table of events based on archaeological data showing metallurgical development in a time span gives the age for the first extraction of lead as 9000-7000 B.C., copper as 3800 B.C. and iron implements from extracted metal as 2500 B.C. This dating for iron, however, is much earlier than what is currently accepted.

Radiocarbon Dating

Literary evidence can be used for dating only from the middle of the first millennium B.C. Archaeological stratification provides a relative sequence of cultures, but it requires radiocarbon dating to give an absolute chronological framework for them. In India, extensive studies on the application of radiocarbon dating to archaeology were made during the last decade and a half, as a result of which reasonably consistent chronologies for various cultures have been established. Indian work has used 5730 year as the half life value with no corrections applied, to study radiocarbon chronologies of the various cultures. Thermoluminescence dating in India is only in its infancy and coming years may yield data to complement radiocarbon dates.

Archaeological Findings

Based on a study covering various sites in India using radiocarbon technique, it has now been possible to establish approximate chronologies for the different cultures which prevailed in this country. In India, the last deglaciation started around 15,000 B.C. in the Kashmir area and the Holocene period could be placed around 11,000-10,000 B.C.

The available C dates show that the Mesolithic culture in India started around c 6000 B.C. The Neolithic cultures are marked by a settled life, agriculture, hand-made pottery and ground stone tools. These cultures have been distinguished under three phases. Phase I, where no metal is found, has been assigned to c 2450-1900 B.C. In Phase II, copper makes its first appearance along with mud and daub houses with a time bracket of c 1900-1350 B.C. Phase III, c 1400-1000 B.C., is marked by more extensive use of metals. Iron appears in the beginning of the first millennium B.C. The Harappa culture of Indus civilisation appears suddenly, without a visible slow evolution. Though Harappa has not been fully C dated so far, and its script not yet deciphered, the beginning of Harappa culture has been assigned to c 2300-2400 B.C. The Chalcolithic cultures, which are customarily treated separately from Harappa are characterised by the use of copper and stone tools. The C dates extend over a wide range from c 2000 to 800 B.C. With the advent of iron, the Gangetic plains were colonised, leading to urbanisation. In the North, the Painted Grey Ware (PGW) and the Northern Black Polished Ware (NBPW) were the early Iron Age cultures. The C dates for PGW fall within a bracket of c 800-300 B.C. except for one with 1025$\pm$110, while the NBPW sites cover c 600-50 B.C.

The recent findings in Indian archaeology with radiocarbon

dating, reveal that, in general, from the Mesolithic to the Iron Age, Indian culture started later than the West Asian counterparts. The advent of copper technology in India is about 1000 years younger, and that of iron, a few hundred years later than West Asia. Whether Indian technologies were indigenous or borrowed, their, dates are certainly later than West Asia. It has been said that the earliest Iron Age sites in the Ganges valley, which date to the early centuries of the first millennium B.C., suggest that the iron industry was in an experimental stage and slowly developed in the south around 1100-900 B.C. In the light of modern archaeological evidence, it can be definitely stated that iron was in use in India around the first millennium B.C.

Ancient Glory

A period of several centuries elapsed between the composition of the Rg Veda and the Age of the Buddha. This was the period relating to one of the greatest of India's Epics, Mahabharata. During this time, the Aryans shifted eastward down the Ganges and the centre of culture and political power shifted to the Kuru capital, Hastinapura. An event of great significance was the battle of Kurukshetra, near Delhi. Vivid descriptions are to be found of this epic battle, where extensive use was made or iron swords, spears, axes and arrows. The iron mace of Bhima received special mention. It has been started that the blind king Dhritarastra broke to pieces, in great agony, a life size iron statue of Bhima. Later, it has been told that the Yadu Dynasty came to an end through a big iron mace. Such accounts give concrete proof of the wide use of iron in weaponry in days of yore. Though tradition places the Mahabharata War around the third millennium B.C. probably the war took place around the beginning of the ninth century B.C.

From then, iron occupied an important place in wars down the corridors of time. Herodotus, the ancient Greek historian, has testified that the Indians in the Persian army of Xerxes, used cane arrows tipped with iron in the battle of Thermopylae about 480 B.C. Ktesias has mentioned swords made of Indian steel presented to him at the Persian court. According to Quintus Curtis, Alexander the Great defeated Porus of Taxila in 326 B.C. and received from him a gift of 100 talents of steel. Indian steel was much prized by the Romans.

After Alexander's invasion, Chandragupta Maurya rose to power around 324 B.C. and drove out the Greek garrisons. He was the chief architect of the greatest of India's ancient empires. Megasthenes, who visited India in 302 B.C. as emperor Seleucus Nicator's ambassador at the Maurya court of Pataliputra, has left an authoritative account of the glory of the Mauryan empire. Asoka, third in the Mauryan line, ascended the throne in 269 B.C. and was the noblest ruler India has known, and indeed one of the greatest kings of the world. The National Flag of India has, at its centre, the design of the Wheel of the Law, which appears on the abacus of the Sarnath Lion Capital of Asoka, an adaption of which forms the State emblem of India. Innumerable inscriptions of Asoka are the oldest surviving Indian written documents of any historic significance. They consist of edicts engraved on rocks and pillars, scattered all over India. These inscriptions on stone required the deployment of very sharp chisels, and it could, therefore, be inferred that the technique of producing hard tools for rock cutting was well known. Asoka built a monastery first at Bodh-Gaya in the third century B.C. Recent excavations of the foundations at this place have yielded metals, including iron clamps, coins and precious stones.

A very interesting specimen of iron slag has also been unearthed and is now preserved in the Calcutta Museum. The presence of slag is an ancient archaeological evidence for the manufacture of iron at that time. Iron articles such as spikes, nails and pieces were found in the monastery at Piprahwa, about 20 kilometres from the ruins at Kapilavastu near the Nepal border. These may be regarded as specimens of iron from the first or second century A.D. A large number of pre-historic iron implements such as swords, daggers, tridents, spears, arrows, spades and rods have been unearthed during the excavation of the burial site at Adittanallur, in the then Pandyan kingdom at the extreme tip of Tamil Nadu. The age of the burial site cannot be ascertained with any approximation, but should be related to an age when the custom of cremation was not introduced into southern India. However, from the reference to the Pandyan kingdom found in an Asoka inscription of 250

B.C. in northern India and the fact that the Pandyan kings had a very long reign, it may be inferred that urn burial was prevalent during the earliest days of these kings.

Delhi Pillar

Beginning from the early centuries of the Christian era, iron has found extensive uses. The famous Iron Pillar of Meharauli near the Qutub Minar in Delhi has attracted considerable attention in view of its uniqueness and the state of its excellent preservation over fifteen centuries. It is a memorial to a king probably Chandragupta II, who ruled India between 376 and 415 A.D. It is believed that the pillar was erected at some other place, variously described as Ambala or Mathura, and subsequently moved to its present site. This view is supported by the comparatively new structure of the platform on which the pillar is erected as also by a later inscription on the pillar. The pillar is over seven metres tall and weighs about seven tonnes. It is said that it would not have been possible to produce an object of this magnitude with the then existing state of knowledge of the working of metals even in Europe.

An analysis of the iron has revealed that it contains 99.72 per cent Fe., 0.80 per cent C 0.046 per cent Si, 0.006 per cent S and 0.114 per cent P, with no manganese. The analysis shows the pillar to be of pure wrought iron. Several studies have been made on the reasons for the non-rusting of the pillar. The absence of corrosion has been attributed to the rather high phosphorus and the negligible sulphur and manganese content of the metal as also the dry weather conditions and the lack of corrosive compounds in the environs of Delhi. The nature of the dense iron of the pillar, the good forge welding employed and the less cold-worked condition also contributed to corrosion resistance. Moreover, a thin film of iron oxide has been observed on the surface, which protected the metal from the onslaught of the elements. This pillar of high quality wrought iron represents the level of macro technology of the time. It has to be inferred that the Megalithic ironsmiths had evolved a technique for forging small blooms of iron into a huge piece. It is likely that the descendants of the Megalithic ironsmiths, who might have been absorbed into the then emerging sub-caste spectrum of Indian society, had contributed their technical expertise, perhaps under royal patronage, to the forging of the famous historic metal-vestige of the world. However, there is need to look at the still unsolved problem of the origin and the production of the Delhi pillar.

Two less famous, but equally interesting, iron pillars are to be found at Mandu, near Dhar, the ancient capital of Malwa, and at Mount Abu in Rajasthan. The Dhar pillar dates to the twelfth century. It was probably intact till 1304 A.D. but now lies broken in three pieces. It seems to be more than double the height of the Delhi pillar. The pillar at Mount Abu stands at the temple of Achaleswar, built in 1412 A.D. and has a trident on its top. The iron of this pillar awaits closer examination. Iron beams of wrought iron as also iron pieces have been used in the construction of the famous Sun Temple at Konarak in Orissa. The forging of the beams is not so perfect as in the case of the Delhi pillar and these have rusted extensively. The iron beams at Konarak date to the 13th century A.D. which would show that the making of wrought iron and forging it into large pieces has been in practice for over eight centuries.

2

Location of Iron Mines

There is hardly any iron bearing region in India where we do not find old working and slag heaps. On the basis of these remnants a list of regions has been drawn up where iron was worked and sources open to smelters during this period:

1. *Sind*: Ranikot area (magnetite, red and brown haematite).
2. *Baluchistan*: Bolan area (clay iron-stone)
3. *Northwest Frontier Zone*: Bajaur (black magnetic iron sand), Bannu (earth haematite), Hazara (magnetite and haematite) and Waziri hills (limonite).
4. *Kashmir*: Different parts of the state (magnetite, haematite, limonite and other sedimentary ores).
5. *Salt Range and Kot Kerana Hills*: Earthy haematite.
6. *Punjab and U.P. Himalayas*: Kangra (magnetic and micaceous), Shele near Shimla (magnetic and Micaceous) Mandi (magnetic and haematite micaceous schists), Sirmur (magnetite, probably mixed with specular ore), Kulu, Almorah, Garhwal (red and brown haematite);
7. *Patiala*: Brown haematite and Magnetite;
8. *Rajasthan*: Alwar, Jaipur, Udaipur, Ajmer, Bharatpur, Bundi, Jodhpur, Kota (basically haematite, magnetite, limonite);
9. *Gujrat*: Surat, Panchmahal, Rewa Kantha, Ahmedabad, Kutch, Kathiawar (lateritic ore in Kutch,-Kathiawar; ore unspecified in the res);
10. *Central India*: Almost the whole of it; lateritic ore in Bhilas, Ujjain, Shajapur, Shivpuri, Mandasore, the ore

of the Vindhyan system principally in Mandasore and that of the Gwalior series in Gwalior; the ore of the Bijawar, series in Indore. Jhabua and Dhar; also, haematite in Lalitpur, Hoshangabad, Narsingpur, Nimar; important deposits in Bastar, Chanda, Durg; other areas; Jubbulpur, Bilaspur, Banda, Mandla, Rewa, Bundelkhand;

11. *Maharashtra*: Konkan laterite; Ratnagiri, Kolhapur, Mahabaleshwar, Goa;
12. *Mysore*: Laterite of south Konkan; haematite—quartzites and titaniferous ores; magnetic sand; areas: Tumkur, Mysore, Chitaldurg, Kadur, Kolar, Shimoga;
13. *Kerala*: Magnetite and laterite, also black magnetic sand; areas: Malabar and Travancore;
14. *Madras*: Magnetite, haematite, laterite; specular ores; areas: Tinnevally, Madura, Balem, Trichinopoly, Coimbatore, Nilgiris, Pudukottai, Chingleput, north and south.
15. *Andhra*: magnetite haematite, limonite, etc., areas: Cuddapah-Kurnool, Guntur, Belbellary, Nellore, West Godavari, Krishna, Vizagapatnam, Hyderabad;
16. *West Bengal*: Different ore-types including the magnetic haematite of Borai-Keonjhar,—Talcher, Sambalpur, Mayurbhanj sand in the western and northern parts of the state;
17. *Assam*: Clay iron-stone and an impure limonite in upper Assam; titaniferous magnetic iron sand in the Khasi Jaintia hills.
18. *Bihar*: Various ore-types in the entire Chhotanagpur block.
19. *U.P.*: (except Nainital-Almorah-Garhwal): magnetic and magnetite sand in Mirzapur.

During 18-19th century smelting of iron was being practised in almost all parts of the country but the industry was more prosperous in the present states, of Assam, Bengal, Madhya Pradesh Andhra Pradesh, Tamil Nadu and Orissa.

Iron Industry in Assam

From 15th century Assam was the seat of iron smelting;

Tirugaon and Hattigar being the renowned centres of iron manufacture. The iron ore used was the ferruginous sand. In the early of 19th century more than 3000 ironsmiths were actively engaged in the industry, however the industry went into oblivion in the middle of 19th century. Cannons and Small arms were the cheif products of this industry. In Khasi and Jaintia hills a large number of smelting furnaces were operating at the time and local weathered rock was mined and produced at Khasi was superior to the British iron and iron worth Rs. 55,000 to Rs. 70,000 was exported to neighbouring areas. Walters, Cracroft and Yule describle the smelting operation in the Khasi Hills.

Iron Industry in Bengal

Birbhum and Burdwan districts were the centres of flourishing iron industry. The ore available locally contained 28 to 59 per cent iron and 1.5 per cent phosphorus pentaoxide. Burdwan district had very rich deposits of coal also. Indigenous industry was better developed in Birbhum district than elsewhere in India. Here the furnace were comparatively large and working was more efficient. Some of the furnaces here also produced iron in Molten condition and steel making process was a second operation which resembled puddling process. In 1852 there were 70 indigenous furnaces in this district. Each furnace produced about 25 maunds of iron in one smelting which lasted 4 complete days and nights. Cost of 25 maunds of iron was Rs. 17.

Iron Industry in Madhya Pradesh

During 18th-19th century iron was smelted in various districts of Madhya Pradesh viz. Jabbulpur, Mandla, Bhandara, Raipur, and Narsinghpur; Tendukhera and Omarpani were the most important sites. The ores of Omarpani producing excellent quality of steel were calcareous in nature and were worked upto 40-50 feet deep. In 1855, about 80 furnaces were in operation, producing about 40 tons iron per year. The rates for crude and refined iron were one rupee a maund and two rupees a maund respectively. Tendukhera iron was used in the suspension bridge over the river Bias in 1830.

Iron Industry in Andhra Pradesh

Of the various iron smelting centres in Andhra Pradesh, Konasamudram produced excellent quality of steel known as Wootz or Nirmal. The daily output of a furnace was about 50 seers, valued Rs. 37. Persians from Isphan used to purchase steel from here which was used for making Damascus blades in Persia and elsewhere.

Iron Industry in Tamil Nadu

Iron industry in the Salem district of Tamil Nadu was well developed during 18-19th century. Here magnetic ore associated with quartz was used in manufacturing wootz. For making steel either of the following two methods, was adopted at each smelting centre, (a) by carburisation of wrought iron in crucible, and (b) decarburisation of cast iron. Furnaces for smelting iron were also in operation at Lakshmipuram (West Godavari district) and adjoining provinces.

Iron Industry in Orissa

In Orissa, iron industry was quite thriving at Balasore, Talcher, Sambalpur etc. In 1872 more than 3000 persons were engaged in iron smelting in Sambalpur and the annual output was nearly 1000 maunds.

3

Mining

Indian miners could recognize by the nuance of the soil and the type of flora, a suitable site for procuring iron ore: Very often the ore was found on the surface.

Open Cast Mining

Mining was generally restricted to surface out crops. With the help of a pick axe trenches were made with stairs on sides. Ores were dug out and taken out of the trench filled in cane baskets.

Shaft Mining

Mining was done by narrow adits or slopes driven from the strike of the ore body Openings of the mines 1 metre (approx.) in diameter were made, just big enough for a person to crawl inside with the support of ropes. Openings were made with the help of pick axe. Eliptical shafts were used for giving roof supports which reflects the sophistication of mining technology of that period. Timbering was also resorted to. Small pillars of ore were left to support the ground and in some cases timbers 15-18 cm in diameter were used to support the stoped out areas.

The labourers while at work sat upon his heels with the lamp upon his head, holding a chisel in his left hand, struck with a hammer held in the right hand and collected the chipped fragments in the basket held upon his knees. The bamboo baskets of ore were passed along a chain of persons, usually 4 to 12 in number, who sat upon their haunches in the galleries or stood on the niches cut in the shafts; when the whole reached the foremost man, they moved upwards, and again ranged themselves, and this was continued until the mouth of the mine was reached.

Tools of the Miners

Pick Axe: The handle was made of a piece of wood with a knob at one end, into which a piece of hard iron was thrust and sharpened at the point.

Iron Hammer: Special type of hammers made of iron 10 inches long were used (see photo 1A).

Chisels: Iron chisels pointed at one end 3 inches approx in length were used for following the strike of the ore body (Photo 2 B).

Spade: Spade made of iron was used to collect the ore from the ground (Photo 2 C).

Pitchers: There are evidences to show that earthen pitchers were used as it has been found in the old workings (See photo 2).

4

Smelting

Benefication of iron was done to enrich the metal content and also to improve the physical characteristics of the ore charged in the furnace. It was achieved by crushing sizing and agglomeration of resultant fines by putting a large hard stone on the ground on which they laid the ores, and then either with a stone or hammer, pulverized the ore at places with the help of water mill which was washed in water, which had a profound effect by decreasing the silica and alumina content. Indian metallurgists knew that the preparation of sized iron ore burden could achieve.

(a) Reduction in Charcoal rate by reducing the metallurgical load on the blast furnace by better utilization of available heat and

(b) Increased productivity and better quality of Iron.

Smelting

Smelting of the beneficated ore was done in a blast furnace, which was one of the first steps in the transition between the original method of reducing the ore in an open fire and its reduction in a modern blast furnace. Doubtless the Indian furnace was the descendent of such a structure. Blast furnace consisted essentially of a vertical cylindrical furnace, built of clay, circular in shape ringed by bellows (wind box) or air chamber from which air blast entered the cupola with fuel and the materials making the charge. There were mainly two types of blast furnace used in India:

1. Fosse Type
2. Aerial Type

Furnaces

Their smelting furnaces, though rude in appearance, are nevertheless very exact in their interior proportions, and it has often surprised me to see men who are unquestionably ignorant of their principle, construct them with precision, in so simple a manner; their unit of measure is the breadth of a middle sized man's finger; 24 of which constitute their large and 20 their small cubit; thus there is a constant ratio of 6 to 5 prevailing throughout these furnaces, nor is it of the least consequence, that their dimensions are larger or smaller, so long as all the parts are in the same proportion; the length of these measures is on an average 19-20 English inches for the large cubit, and 16 English inches for the small one.

As they have no standard measure their fingers, their span, and their arm are substituted by which a piece of stick is measured which they use in practice; neither is the division of the cubit necessary though the large one supposed to be divided into six parts and the small one into five, of four fingers each—as the measurement is invariably ascertained by their fingers; the length of these parts is on an average 3.20 English inches.

Geometrical Construction of the Furnace

To construct the outlines of the furnace geometrically (Plate 3a figs. 1 and 2) rule an indefinite line A.B. which suppose equal to a large cubit of 24 digits or 19-20 English inches, and divide it into 6 parts; at C erect a perpendicular, then from C to E set off 6 parts and it will mark the central point of the greatest bulge, and consequently the point of greatest heat; next, from E to F set off 6 more points, and it will mark the point of cremation; then again from F to G, 6 parts more, will mark the line, where it is necessary to recharge the furnace, after the burden has sunk thus low, and from G to D—two parts more; will give the perpendicular height of the furnace, in 20 parts equal to 5 feet 4 inches of English measure.

To complete the figure, rule lines parallel to the base, through the points E, F, G, and D, and from D, fig 1, set off three parts to the left hand for the top; bisect it at J, bisect also the bottom at H, J, right angled at K, and it will be the oblique axis of the furnace (fig 1, K-J) bisecting all the parallels corresponding with CD (Fig 2)—then make the parallels AB six parts,

—E six parts, F five parts, and D three parts; rule lines through all these points, and the geometrical outline will be completed, the sum of the parallels in parts, corresponding with those of the perpendicular.

Practical Construction of the Furnace

To construct it practically—dig a fosse 3 feet deep (Plate 5, fig 2) the semicircular part of which contains the furnace B, the walls CCC being composed of unburnt bricks of large dimensions; the first structure is rude, preserving only an approximation to the required form, the interior being afterwards cut away; a large stone capable of containing heat is placed at the bottom; and in this state it is suffered to remain until thoroughly dry; the next operation is performed by a more skilful artist who cuts away the interior, and plasters it with clay, using the measures above described to adjust its dimensions; he first finishes the top, and from the centre of the back part of it, he drops a plummet, to ascertain the spot where the centre of the front part of the stone is to rest; this plummet line corresponds with the perpendicular CD of the geometrical figures 1 and 2—and thus he obtains not only the required obliquity of the furnace, but the points most essential for the adjustment of all the rest.

When the furnace is thus far prepared it is again suffered to dry, and in the mean-time, other appendages called by the Indian smelters Gudaira, Pachar, Garrairi, and Akaira (names which have no synonimes in the English language), are constructed; the Akaira in particular is a most extraordinary implement, (Plate 3 figs 4 and 5; and Plate 4 fig 1+); externally viewed it is a clumsy mass of clay enveloping the wind tubes (Plate 3 fig 9) but when it is considered that the complete fusion of this mass, and the perfect completion of the smelting process must be simultaneous results, the implement becomes the most important of all the appendages; thus for instance if it is too small, or too large, its effect will immediately be perceived; in the former case the masset of crude iron will be full of impurity, and in the latter the iron will be consumed, and if it cracks during the operation of smelting, there is no remedy for such an accident—short of dismantling the furnace and commencing the work again.

I found after numerous experiments that its mean length should be 4½ parts, its mean breadth 3 parts, and its mean thickness 1½ parts and it is somewhat remarkable that the product of these dimensions, should exactly equal a twentieth part of the cubic contents of the furnace when fitted for use; this coincidence may arise from the peculiar nature of the clay of·Tendukaira, the ingredients of which are well assorted, but the rule will nevertheless apply generally—because clay by admitxure is susceptible of being rendered amenable to rule; and therefore this implement will be found in all Indian furnaces to have (or by tempering the clay may be made to have) the same corresponding dimensions.

The Guddaira is a wedge of clay used to adjust the vertical position of the Akaira when placed in the furnace; and the pachar is an oblong plate of clay, used in walling up the orifice after the Akaira is placed, and adjusted; these figures and dimensions are given in Plate 3 —figs 7 and 8; the Gurairy (Plate 3, fig 6) is a convex plate of clay; perforated with holes and used as a grate—through which the scoria are drawn off.

When the appendages are ready, and the furnace thoroughly dry, it is prepared for use in the following manner. The front part is walled up from the top to the line SS—Akaira to the top (Plate 3 figs 1 and 3) which line is ascertained by the small cubit; one end being placed at C, the other will measure CB and CS (fig 1)—the grate is next put in, its lower edge resting upon the edge of the stone; and the space is filled—with a mixture of pulverised cow dung and kodo straw—up to the dotted line (Plate 3 fig 1) upon which is placed the Akaira; its sides being every where 1½ parts distant from the walls of the furnace—as represented in Plate 3 fig 4 and Plate 4 fig 1+; where a, b, c, d, are the walls of the furnace, fig: 5 and 1+the Akaira; the Gudaira or wedge, is next introduced in order to adjust its vertical angle (Plate 3 fig 1) and this being placed satisfactorily, the Pachar is inserted and the whole has then the appearance represented in Plate 3, fig 3, where Nos. 5, 6, 7 and 8, are the Akaira, Gudaira, Pachar, and Garrairi; nothing now remains but to lute the whole with clay, leaving the ends of the wind tubes open to received the bellows.

Construction of the Furnace: (Aerial Type)

The material used for the construction of the native furnaces was the common red potters clay of India (which unless carefully selected was not generally very refractory and will hardly stand heat sufficient to fuse cast iron). However by mixing it with sand and concentrating the heat in the centre of the furnace as much as possible by projecting the blast pipe, and the reduction of the ore was effected before the furnace becomes much more than red hot; and the operation was completed in about a couple of hours. A platform about 24.50 cm. in square and 18.00 cm. thick was first made, with a hole in the centre 32.00 cm. in diameter—a half cylinder or curved piece was then formed also of the red clay 63.75 cm. high, 14.00 cm. thick and 46.00 cm. diameter inside, and also a cone about 74.00 cm. thick of the same height and the same diameter at bottom and 24.75 cm. at top. When these were quite dry, a little wet clay was put round the hole in the platform, the half cylinder placed upon it and the spent front built up with clods of clay. The inside part was plastered (thickness 7.00 cm.) until a hollow cylinder was produced, about 81.50 cm. deep and 32.00 cm. in diameter inside and a thickness of about 21.00 cm. When nearly dry, an arch was cut out in front at bottom, and about 32.00 cm. high, to form the door of the furnace. The cone was then placed on the top, and inside plastered with clay to correspond with the bottom part, and the neck or throat reduced in the same way to about 5″ (18.00 cm.) in diameter. The upper part of a chatty with the neck was then placed inverted on the apex of the cone, to form a funnel so as to conduct the charge into the throat. The Chatty and the whole of the outside of the furnace was then plastered over with clay, about (7.00 cm.) thickness. When finished the height inside from the bottom to the neck was about (162.84 cm.) and the whole operation took about a week to finish before it was quite dry. The blast pipe was a cylinder of dry clay (49.50) long, about (14.00 cm.) thick, pierced with a hole of one inch in diameter. It was introduced into the furnace at the bottom of the door with its point about the centre of the furnace and about 5″ above the bottom. The door was then closed with a tile of dried clay, and the outside was built up and seamed with wet clay, and plastered over; a layer gf charcoal dust about (7.00 cm.) thick having been first placed at the bottom of the furnace to prevent the reduced oxide adhering to it.

The blast pipe was connected to the bellows of the whole goat skin.

The bellows are as singular in their construction as the Akaira, and are worked by the hand; they are made of a single goat skin, the dimensions of which ought to be 7 parts in breadth when doubled, and 8 parts in length; such proportions being required for circular bellows of 5 parts diameter, and which when worked by a man of ordinary strength will rise 6 parts in height—having 11 1/4 circular folds; the wooden nozzle through which the blast is conveyed into the furnace surfaces the Akaira in its complex nature, and so little is its principle understood—that the art of making it was once lost at Tendukaira, and was again restored by the smelters of Katola.

Construction of the Nozzle of the Bellows

To construct its figure geometrically, rule a line AB equal 3 parts (Plate 4 fig 2) divide it into four, giving one of those divisions to each of the legs, and two for the space in the centre; set off a perpendicular from C to D equal 3 parts; bisect it and the middle point will mark the apex of the central angle; then through the point D rule a line parallel to AB and from it as a centre set off each way 3/4 of a part making together 1½ parts; divide it also into four, giving one of each to the legs, and two for the space in the centre as before; and then by ruling lines to connect all these points, the outline will be complete; the exterior of the implement is plain but the interior is complex and cannot be described except by a reference to plate 4 fig 3 which represents it, divided in the middle, to shew its internal structure.

This curious appendage is fastened to the bellows by leathern thongs, and the blast is forced through it at an angle of 24° but when it is luted to the wind tubes of Akaira, the blast enters the furnace at an angle of 12°, both vertically and horizontally—because those tubes are placed so as to reduce that angle; Plate 4 fig 1+ represents the whole apparatus luted together and placed in the furnace the walls of which are marked a,b,c,d, and it exhibits at one view, the whole of the mechanism of this complex machinery; the furnace when closed up with clay, and the bellows luted in, is represented entire in Plates 5 and 6; the dotted lines shewing the chimney—A the outer walls, B, a mound of earth to strengthen the walls, C an upper chimney of moveable bricks, D planks laid across the trench to support the bellows and the man who works them, E a stone supporting one end of the plank, F forked branches supporting an iron bar on which the other end of the planks rests and G a simple apparatus for preventing the bellows from rising from the planks when they are worked.

The above description is not founded on theoretical conclusions; the measurements given are derived from taking the mean of several and the results were proved in furnaces under my own superintendance; the coincidences of the several parts are very striking, thus for instance, the perpendicular and parallel lines of the geometrical outline are equal in quantity (Plate 3, fig 2); and the top, bulge and bottom being 3, 6 and 4½ parts respectively, shew that the furnace is exactly constructed, and that it corresponds well enough with the most regular furnaces of Europe (Plate 3, fig 1); it is also curious, though perhaps of no importance to observe, that the mean of those numbers, being squared and multiplied by the terms of the perpendicular or axis, give the cubic area of the furnace, and shew that it is twenty times larger than the cubic content, of the Akaira; the angle of the blast is also worthy of notice, as well as the simplicity by which both it and the obliquity of the furnace is obtained; all these serve to shew that the original plan of this singular furnace must have been the work of advanced intelligence, and that is geometrical proportions have been preserved by simple measures; hence though its original form may be changed by caprice or ignorance, its principle never can be lost so long as hands and fingers remain.

Refineries

The refinery is as rude in its appearance, and as novel in its construction as the furnace, to which it seems to have been purposely adapted: two refineries being required for one smelting furnace; to construct it they use the small cubit of 20 digits, or what is still more available, the space or distance between the tips of the thumb and little finger of a middle sized man when extended without force, two such spans being considered equal to the cubit; the first process is to arrange a number of square unburnt bricks, as in the ground plan (Plate 5 fig 1) in which a.a.a.a. are the walls—A is the chimney—B the refining furnace, C the seat of the refiner, and D the anvil; see also (fig 2) for a side view of it—divided in the middle for the purpose of shewing the interior structure, in which E is a piece of crude iron under the process of decarbonization; the dimensions of the chimney are not material—but it is usually about one cubit broad, one deep and sex in length; the oval part where the operator sits is altogether a fanciful appendage, being merely a mound of earth in which a log of wood is inserted for receiving the anvil—and its elevation serves the further purpose of giving the workmen a purchase in using the hammer; when the walls of the chimney are finished, the top is covered with unburnt bricks of an oval shape, flat below and convex above and these are luted together by a plastering of clay—fig. 3 is a front view—shewing the opening of the furnace and Plate 8 exhibits the refinery complete, with the refiner at work on his seat, the bellows-man plying the bellows, and various implements lying about—A the outside of the chimney—B a mound of earth to strengthen its wall—C the refining furnace—D a piece of crude iron undergoing the process of decarbonization (in dotted lines) —E the bellowsman plying the bellows—F the refiner with an iron spike in his hand regulating the operation (the dotted lines shewing the interior of the furnace)—G a thick plate of iron placed at the bottom of the refinery (in dotted lines)—H a fosse for the hammerman—I the anvil—K implements, and—L a heap of charcoal.

The furnace of the refinery is the only part which requires skill in its construction, and this is usually done by the operator himself; its geometrical outline is represented (Plate 7 fig 4) and its construction is as follows. Rule a line AB equal five parts, divide it into six—set off four of these divisions for the top—let fall a perpendicular [from the centre C—set off three divisions from C to D for the depth—rule a line through D—parallel to AB and make it two divisions, now rule the outline—bisect the perpendicular and the centre parallel will be equal to three divisions.

The centre parallel is the most important part of the furnace and next to it is the accurate adjustment of the angle of the blast; I have frequently seen the Indian refiners obliged to discontinue their work on account of some error in this point; the usual measure for the former is the span abovementioned applied both longitudinally and transversely as in Plate 7 fig 1, B where a ground plan of the furnace is exhibited—the inner circle of which corresponds with the centre parallel of fig 4—this measure never differs much from 8 inches, and that quantity may be assumed as a fair average; the outer circle b of the same figure is indefinite—the space between the two—being merely a slope, chamfering from the inner edge, and gradually expanding, until it is lost in the sides of the furnace, so that in fact it is reverberatory; with regard to the blast, it is absolutely necessary that it should be directed, at an angle of about 12°, upon the opposite edge of the inner circle or to the point c fig 1, B; the natives have no instruments to enable them to do this exactly but the working of the furnace soon tells them where there is an error and they know well enough how to correct it;—the bellows resemble those of the smelting furnace but instead of the wooden nozzle, they are furnished with long iron tubes—as in Plate 7 fig 5 which are so placed, that the angle of the blast thrown through them is 24°, the same as that of the wooden nozzle.

Smelting Furnace

Plate 9; figs 1 and 2 represent the front and back view of a small circular smelting furnace, which is very common in India—its measurement may be taken from the scale of the Plate either in parts or inches, the bellows are the same as fig 5; Plate 7, and the form of the interior or chimney is exhibited by dotted lines; figs 3 and 4 of the same Plate shew another description of refinery used chiefly for decarbonizing large masses for the manufacture of auvils and c worked by two pairs of bellows—this refinery might be more extensively applied; such as for the manufacture of adletrees or other heavy work. Fig. 5 is a small field black-smith's forge, constructed of the same kind of oval bricks, as those which are used for covering in the refinery, and luted together by clay; this apparatus may be constructed in half an hour; and is a useful field-smithy; fig 6 is a tube of clay, used in the refinery to preserve the ends of the bellows—fig 7 is a tube of the same kind used in the small circular furnaces.

Mode of Smelting and Refining

In the process of their manufacture the Indian smelters use charcoal only; the ore is broken into pieces about the size of a walnut, but it is not washed, nor is it roated although it is known to contain a large quantity of sulphur which might be dissipated by that method; they commence by filling the chimney of the furnace with charcoal which they suffer to burn until all moisture is expelled; they then throw in a small basket of ore, and upon it a larger one of charcoal, after which the burden is allowed to sink as low as the line G (Plate 3: figs 1 and 2) when it is again charged—and afterwards ore and charcoal are alternately given in the same proportions until the operation is complete; the scoria begin to flow within the space of an hour, and by that time, it is known whether the furnace will work well or ill—the scoria being a sure indication; it is let out by piercing the grate with an iron spike, and the orifice is again closed with clay as soon as it is drawn off; the bellows are worked by three men—who take it by turns; and they should be kept constantly playing until the process is completed; the time of which is ascertained by introducing a hooked piece of iron through the wind tubes, into the furnace, which shews how much of the Akaria remains; for as I have shewn before, it is indispensably necessary that this appendage should be totally fused before the operation is complete, and when this is the case, it would be useless to continue longer, because the furnace would cease to work properly; it usually continues 12 hours but much depends on the bellows-man, and also on the working of the furnace.

The metal is never completely melted by this process—the heterogeneous mixture of the ore alone is fused and thrown off in scoria, and the iron being freed from it falls by its superior gravity to the bottom of the furnace, and there coagulates into a mass or masset; it is never very highly carbonized, and sometimes it is partially malleable even in its crude state, when the process is finished, the bellows are removed, the front part of the furnace demolished, and the red hot masset dragged out, and divided by large ades before it has time to cool, hence it happens that the parts of the furnace thus broken up, require daily renewal.

This completes the business of the smelting furnace—the process of decarbonization being performed in the refiner, and

Plate 8: fig. D, represents half a masset—properly placed in a refinery and undergoing that operation; as it becomes decarbonized, it drops into the hollow of the furnace upon the plate of iron G—and when a sufficient quantity has accumulated, it is taken out and hammered into more circular lumps which are seen in every bazar; the charcoal used in the operation is always of hard wood, such as teak (tectona grandis), mowa (fanalatifolia) or bamboo; this is the part of the manufacture in which the Indian manufacturers play tricks; for in the first place they do not allow time for the crude mass to become properly decarbonized and then again they have a most pernicious practice of knocking off corners and small pieces from the masset in its crude state into the decarbonizing fluid, and instead of waiting patiently until the whole of the masset is decarbonized they often thrown in large lumps at the end of the process, mixing all these crude pieces with the other so that the cheat cannot be detected except on trial; thus they not only shorten the time of the operation, and thereby use less fuel, but contrive by this nefarious practice to sell a large portion of their crude iron at the same price as the malleable; they are also very sparing in their hammering lest they should force out too much of the Vitreous Oxide and thereby reduce the weight; so that upon the whole there seem to be causes which have justly affected the reputation of Indian iron—but as they are repairable errors, they ought rather to be placed to the account of the perversity of Indian habits—than arrayed against that fair repute to which the Indian metal under different management might lay claim.

5

Steel

The method for making steel was to convert wrought iron in crucibles, resembling modern cementation process.

Wrought Iron

Whether smelted in open hearths or in small and large blast furnaces, product was wrought iron. Indigenous method of iron production during the period was that of direct reduction. Indians took the advantage of the fact that despite the high melting point of iron (1540°C) iron oxide could be reduced to metal at a low temperature of 800°C. At a temperature of 1100-1500°C, this reduced iron flows together forming a semi-fused porous mass called as bloom. The bloom can be easily forged and worked. Flux was not used, save in the case where the ore was self fluxing. This had the disadvantage that lot of iron oxide passed into slag but helped in removal of phosphorus if present in the ore of fuel and also rendered the slag very fusible.

This low temperature smelting avoided the reduction of any other metal present in the ore and also diminished the danger of carborization of iron. The product was slag bearing wrought iron. It was fibrous in nature and could be easily welded and forged. Metal was very pure and no other metal including manganese was present. Sulphur was very low and consequently metal had high resistence to corrosion. For chemical composition of wrought iron refer table.

However, it must be admitted that the yield from best ore rarely exceeded 50 per cent of iron in the ore. Ordinarily the yield were of the order of only 20 per cent and the remaining was lost in the slag. Rich iron slags littered all over the country prove the point. (See Table 5.1).

TABLE 5.1: **Analysis of Iron Slags from Various Smelting Centres.**

Description of sample	*SiO_2*	*Al_2O_3*	*FeO*	*CaO*	*Loss on Ignition*	*MnO*	*MgO*	*S*	*P_2O_5*
Rewakanta Bombay	53.64	5.39	28.96	10.49	1.52				
Madhya Pradesh	10.33	1.85	8.13 +	2.49		0.23	1.07	0.03	0.35
Rajdoha			73.95						

TABLE 5.2: **Analysis of Iron Ore by Harris.**

Ferric Oxide	(Fe_2O_3)	69.55
Ferrous Oxide	(FeO)	19.50
Silica	(SiO_2)	5.83
Manganous oxido	(IInO)	0.22
Alumina	(Al_2O_3)	0.51
Lime	(CaO)	0.36
Magnesia	(MgO)	Trace
Sulpher	(S)	0.02
Phosphoric anhydride	(P_2O_5)	0.03
Moisture at 100°C		0.60
Combined water and loss		3.28
Total		100.00

TABLE 5.3: **Percentage Composition of some Ancient and 19th Century Iron Objects**

Description of the specimen	*Fe*	*Carbon*			*Mn*	*N*	*S*	*Si*	*P*
		graphitic		*combined*					
1. Delhi Iron Pillar 5th Cent. A.D.	99.768	—	0.23	—	—	0.0065	tr	0.026	0.180
2. Konark Iron 1250 A.D.	—	—	0.11	—	—	—	0.024	0.100	0.015
3. Rajdoha crude iron 19th cent. A.D.	98.180	0.66	—	—	0.013	—	0.005	1.110	0.028
4. Rajdoha Hammered & reheated iron, 19th cent. A.D.	99.947	—	—	0.03	—	—	—	0.010	0.030
5. Modern Tribal Smithy iron from Koraput Orissa	—	—	0.23	—	—	—	0.006	0.100	0.015

It is the power of wrought iron to resist outside atmospheric influences better than steel that has caused the production in indigenous wrought iron to survive side by side with modern steel works.

Pocess of Making Wootz Steel

Indian steel generally referred to as wootz was made by a method resembling the modern cementation process or crucible process. Small pieces of wrought iron weighing 1/2 to 2 lbs were packed into crucibles with 1/10th of their weight of chopped dry wood and leaves of soecific plants e.g. Avaram (Cassia auriculata). Crucibles were sealed with clay and up to two dozen of such crucibles were staked into the furnace which was maintained at high temperature by blowing air over charcoal fire. The process was completed within 2½ to 4 hours. The cake so formed was forged at a very low temperature. Composition of wootz steel given by Percy is:

Percy J. Metallurgy: Iron and Steel, London, pp. 254-278.

Combined carbon	1.33%
Uncombined carbon	0.31%
Silicon	0.45%
Sulphur	0.181%
Arsenic	0.037%

This composition agrees with high carbon steel, slowly cooled from above its melting point, which would be nearly 1450°C.

Steel Swords

Before the advent of Britishers and even in the 19th century Indian steel objects were recognized to be of excellent quality which is evident from the fact that Indian steel was in great demand by the cutlers in Europe and the remarkable quality of Indian iron and steel was subject of great discussion among the 18-19th century. British metallurgists. In the so called mutiny of 1857, British Govt. was dazed by the high quality of Indian steel weaponry and ordered their collection. Thousands of swords were turned into non-military tools. It was mentioned that the Indian swords and weapons were hard to break and they could not be cut even with strong sheers. It was impossible to cut thin edges of the Indian blades, indicating that these were the highly carburised areas.

The superiority of the Indian steel could be accounted for by the high quality of Indian iron ores. According to an analysis Indian ores contained 90 per cent iron oxide and a very low amount of manganese, sulphur and phosphorus (Table 5.2). Secondly, the superiority of Indian iron was because of the fact that it was prepared in small quantities at a low heat, in apparatus with a minimum of reducing power, so that the reduction of iron oxides is incomplete and that of other oxides practically nil. Thirdly, in the manufacture of steel by the native methods, the molten steel was allowed to cool down slowly in the crucibles, which is advantageous to the material.

6

Malleable Iron

Malleable iron is a cast alloy consisting principally of iron and carbon which is rendered tough and ductile by a controlled heat-conversion process. Because of its unique metallurgical structure, it has the valuable properties of strength, great toughness, remarkable resistance to heavy and repeated impact, excellent ductility, high resistance to corrosion, and easy machinability. Its versatile castability makes possible sound castings, accurate to pattern, in complicated as well as simple forms, over an extensive range of weights and sizes. The combination of these properties in this one metal has established it as the ideal material for a wide diversity of applications.

The first use of iron is lost in the mists of prehistoric time. It is clear that primitive man learned that heating a type of reddish earth would give him a red hot mass of material which he could pound into shapes of his own design. The reddish earth was iron ore—oxides of iron. The wood fire which heated it formed charcoal. This served as the reducing agent to deoxidize the ore, leaving essentially pure iron which could be 'worked' or 'wrought.'

In later recorded time, the ore was reduced in Catalan hearths, then in their successors, the early stack furnaces. In these, temperatures were sufficient to produce only an incandescent spongy mass of iron. The metal, however, was almost pure chemically, for two principal reasons: First, the temperatures were still below the ranges where chemical combination could occur between the iron and any other elements present—notably the carbon of the charcoal; and, second, at the temperatures attainable the charcoal ash, resulting from combustion of the fuel, sintered or partially fused with the clay and other minerals found in nature with the ore, to form a slag. Some of the slag worked through and out of the hot mass, but because the iron was pasty in texture, much slag remained entrapped. The glowing masses with which the early iron masters worked were known as 'blooms' and their furnaces as 'bloomeries' which in their later built-up forms became the 'high furnaces' and were the forerunners of the modern cupola and blast furnace. The blooms were forged by hammering, to shape and form the articles desired, and to force out as much as possible of the entrapped slag.

Improvements in furnaces and their performance led in time to complications; the iron produced became more difficult to work and when cooled it lacked strength and could easily be broken. Reheating the iron, some of the early ironworkers found, at least partially restored the more ductile conditions they thought of as natural. This rule-of-thumb technique relied on the basic chemical principle that with sufficient temperature carbon will migrate in iron. If the carbon content is higher in the materials or atmosphere the iron than in the iron itself, carbon will enter and migrate through the iron; if the carbon content is higher in the iron, it will migrate toward the surface and out of the iron. The first process—absorption of carbon—is called carburization; the second—removal of carbon—is decarburization. During carburization, carbon combines with iron to form iron carbide, harder and more brittle than pure iron. Decarburization frees carbon from the carbide, restoring the iron to a chemically purer, softer state.

Up to this time the iron obtained by the direct reduction of the original ore could be wrought, for the temperatures had not been high enough to cause any appreciable carburization. Now the higher temperatures provided by improved furnaces had introduced the hardening influence of carbon into ironmaking. As the ore was reduced with charcoal at these more elevated temperatures, the carbon-rich fuel and its gases of combustion supplied the excess carbon that carburized the formerly pure iron of the blooms, making it hard and unworkable. It therefore became necessary to go to the indirect or two-stage process of first producing a carburized hard bloom by reduction of ore, and then by a heat-treatment decarburizing it to prepare it for ultimate forging. When the hard blooms were reheated, preparatory to being formed into desired shapes, much less fuel was needed than that required in the original reduction of the ore, and so the furnace atmosphere was now decarburizing in nature. Hence the carbides were partially broken up, carbon migrated out of the iron, and the iron was chemically repurified, its hardness lowered, and its workability restored.

As the height of the stack furnaces increased, the time of contact between ore and charcoal lengthened, temperatures grew hotter and carbon absorption increased. Ironmasters, who now were making blooms so large that they were unwieldly and had to be pulled apart into pieces of convenient size for forging, were beginning to hope that finally they would be able to melt iron and cast it as other metals were cast. The blooms, they expected, would soften progressively and ultimately liquefy. The time came in the first decade of the fourteenth century when, in some furnace, well before seemingly high enough temperatures were reached, iron actually meltered. Breaking into this furnace for the removal of the expected bloom, the ironworkers found that the fluid metal had run down into furnace hearth, collected into a molten pool, and rid itself of the lighter slag, which floated on the surface. The now well-known capacity of carbon for lowering the melting point of iron had come into play and had produced an iron rich in hard iron carbide, but relatively slag-free and clean.

The introduction of melting had several profoundly important effects. Since working out slag during forging now became almost unnecessary, the indirect process proved to be more efficient and economical than the direct, which involved both a much longer reduction period at lower temperatures to avoid carburization and then laborious forging to drive out the slag. Melting, therefore, was the greatest single influence in the adoption of the indirect process, which became general practice during the seventeenth century. Moreover, once molten iron was available, coupled with experience in the casting of other metals, iron founding was at hand. Castings made of the newly available material had to be confined to a limited field, since they were of a highly carburized iron, and therefore very hard and brittle. The advent of molten iron, too, permitted a time lag in the progression from raw material to finished product, through the device of running the molten metal out of the furnace into shallow, thinly connected troughs to cool into pieces that were easily broken apart. The troughs, when filled, had a most realistic resemblance to a sow with suckling pigs, hence the term 'pig-iron.' For the infant foundry industry of this period, with meagre demand for its brittle, hard castings, relatively little melting stock was necessary. Of the two means for obtaining it—the direct reduction of ore or the melting of pig iron—the second was clearly cheaper and more practical and was adopted where possible, thus contributing further to the establishment of the indirect process. No significant amount of gray iron, malleable iron, or cast steel is made today by a process deriving its basic iron direct through the reduction of ore.

The process of malleablization consisted of melting iron, pouring castings, packing them in pulverized iron ore, and heating them to bright redness for many days. The type of furnace which was available to him is shown in Fig. Lengthy heat-treatment in this temperature range slowly removed the carbon present as iron carbide in the metal as cast. This decarburization was partial to complete, according to the thickness of the casting and the temperature and time of heating. The metal thus produced was almost free from carbon and for the first time provided ferrous castings possessing some toughness and ductility.

At elevated temperatures, the chemical affinity of oxygen for carbon is stronger than its affinity for iron. As carbide gave up carbon the iron ore packed around the castings released oxygen to combine with the freed carbon to form gaseous carbon monoxide and dioxide, which escaped from the furnace. The packing performed three functions: First, it supplied the oxygen for oxidizing the carbon, second, it supported the castings that otherwise would have softened and lost shape under the applied heat, third, it acted as a 'buffer', shielding the castings from direct contact with the hot furnace atmosphere—a contact which in the long period of heating would have caused undue oxidation of iron as well as carbon, creating a heavy scale on the castings and perhaps even destroying thin sections.

This process, with only minor deviations, has been followed in India to the present day in those foundries still producing 'white-heart' malleable iron. To understand this descriptive term, one must remember that this iron is decarburized in the sense that most of the carbon freed from its carbide has been eliminated, though not all the carbide may have been broken up into iron and carbon. The constituents of the annealed metal, therefore, are mainly iron and iron carbide, although some of the freed carbon may also remain. Because of the presence of the iron carbide, the grains of the metal, when under tension, stretch to a limited extent or not at all, and then shear through and across their sections along minute and fairly flat planes which reflect light. These minute reflecting planes give a whitish or grayish color to the reptured surface, and hence its description as white-heart'.

Since decarburization was the basis of the medieval process, only small castings could at first be made malleable by it. Larger ones were liable to be hard and brittle in the centre. Though presentday manufacture of white-heart malleable has advanced beyond these limitations, decarburization continues basic to it. Some graphitization—the precipitation of carbon from iron carbide and its retention in the casting as nodules of free 'temper' carbon does take place in white-heart malleable castings during the heat-treatment, but small castings, say upon a half-inch section, are almost entirely decarburized. Heavier sections consist of a white rim and a core in which some of the carbon remains in the combined form and which contains some nodular graphite. The properties of the rim differ greatly from those of the core, and so the size of casting which can be made successfully in white-heart malleable is still limited.

7

Bar Iron

The mode of smelting iron used by the natives of India, appear to be very much the same from the Himalayas down to Cape Comorin.

The ore principally used was either the common magnetic iron or pounded magnetic iron ore, separated from the ferruginous granite.

The materials used for the native furnaces, was the common red potter's clay of India, which unless carefully selected, was not generally very refractory, and will hardly stand a heat sufficient to fuse cast iron, but by mixing it with sand, and by mixing it with sand, and by contracting the heat in the centre of the furnace as much as possible by a projecting blast pipe, the reduction of the ore was effected before the furnace has become much more than red hot; the operation being completed in about a couple of hours.

In constructing these furnaces, a platform about two feet square and five inches thick is first made, with a hole in the centre nine inches in diameter. A half-cylinder or curved piece is then formed also of the red clay, eighteen inches high, four inches thick, and thirteen inches diameter inside, and the same depth, and also a cone about two inches thick of the same height, and the same diameter at bottom, and seven inches at top. When these are quite dry, a little wet clay being put round the hole in the platform, the half-cylinder is placed upon it, and the open front is built up with clods of clay, and the inside part is plastered for two inches thick until a hollow cylinder is produced, about twentythree inches deep, nine inches in diameter inside, and about six inches thick. When nearly dry, an arch is cut out in front at bottom about nineteen inches high, to form the door of the furnace. The cone is then placed on the top, and the inside plastered with clay to correspond with the bottom part, and the neck or throat reduced in the same way to about five inches diameter. The upper part of a chaty with the neck is then placed inverted on the apex of the cone, to form a funnel to conduct the charge into the throat, and the chaty and the whole of the outside of the furnace is then plastered over with clay about two inches thick, so as to give the appearance of a large sugar loaf enlarged a little at the point. When finished, the highest inside from the bottom to the neck is about three feet ten inches, and the whole takes about a week to finish before it is quite dry.

The blast pipe was a cylinder of dried clay fourteen inches long, and about four inches thick, pierced with a hold of an inch in diameter. It is introduced into the furnace at the bottom of the door, with the point about the centre of the furnace, and about five inches above the bottom. The door is then closed with a tile of dried clay, and the outside is built up and secured with wet clay plastered over, a layer of charcoal dust, about two inches thick, having been first placed at the bottom of the furnace to prevent the reduced oxide adhering to it.

The bellows were two goat skins taken off the animal, by opening the hinder part only. The orifices at the legs are sewn up, and a piece of bamboo is inserted, and tied tightly into the neck of each skin, and these bamboos being inserted into the outer part of the blast pipe, which is made conical for that purpose, the vacant openings are then stopped with wet clay. The open end of the skin is finished by folding the edge of one side, as a flap, about four inches over the other edge, and sewing up the upper and lower corners, so as to leave a part of both flaps open for about nine inches. When the skin is filled with wind, and pressed, the inner flap closes therefore against the outer, and stops the passage. Each skin is managed by one man, who places it in his lap, and squeezes it down with the elbow and lower part of the right arm, grasping at the same time a projecting sort of handle of leather, formed at about the part where the tail of the animal might have been. To enable the blower to fill the skin again with wind, a piece of string is attached to the lower corner of the posterior part, which is tied to a peg driven into the ground about a foot behind the man's elbow, and keeps the skin distended to its full extent as it lays in his lap; and a loop of leather is also fastened to the outer flap, through which the arm is passed, by which the opening into the skin is distended upon raising the elbow; and the skin being pulled out horizontally by the neck in one direction, and the string and the peg in the other; upon pulling up the middle part vertically by the leather grasped in the hand, the skin is opened out into a triangular shape, and fills with wind through the open flap. While squeezing one skin in blowing also, by pressing the hand forward, so as to pull against the string attached to the posterior part, the valve flaps are made to close fairly and properly, so as to allow hardly any wind to escape. The left hand is employed in assisting the right, or in squeezing distended parts of the skin upon the side. It will be observed, that as both the necks of the skins open into the blast pipe, a portion of the wind expelled from one skin passes back again into the other; as they are worked alternately, a defect which might have been remedied in a most simple manner, by attaching little hanging door valves to the ends of the pipes.

A small quantity of charcoal being thrown into the furnace, the fire was introduced, and the blast commenced, and the furnace was filled to the neck with about twenty-six pounds of charcoal. In about half an hour flame issues from the throat, and the fuel beings to sink, at which time the charge is commenced, which consists of ten pounds of charcoal and five pounds of ore, wetted to prevent it running down too fast. The charge is repeated seven times, and the furnace is allowed to burne down, and in about two and half hours as soon as welding heat sparks are seen to issue with the flame, the bellows are removed; the door broken open; and the lump of reduced iron is removed, and cut open while hot with a hatchet, to shew the quality. Four men are required to work one of these furnaces, one being a maistry superintendent, and the other three labourers, and they are able to make about three lumps in a day of twelve hours, but after four days' work, the lining of the furnace is destroyed, and requires renewal.

The lumps of iron were then forged into bars by hand hammers.

8

Smithy

Blacksmithy was an important and commonly met household industry. It was invariably a hereditary occupation of the households concerned. The son worked as an apprentice and received training under his father and the skill was thus carried down from generation to generation.

Tools and Implements

The tools and implements used by the blacksmiths were very simple. A pair of sledge hammers, an anvil, a pair or two of longs, a few small hammers, cold chisels, a furnace with bellows were all that they required for their working: (Plate No. 14).

Furnace: For heating the metal a furnace was necessary. For small works the blacksmith's furnace was commonly used and consisted of a hearth for holding the combustible coke, an opening for leading forced air into the fire and a blower or bellow either of wood or leather for supplying air under slight pressure, through the mouth of the furnace.

Anvil: The anvil was meant [for supporting the work or object while it was struck with the hammer as well as providing space for other forging operations. The weight of anvil even exceeded 3 CWT and stood with the top face about 2 feet from the ground. The height was attained by resting the anvil on a cast iron or wooden base.

Hammer: Two kinds of hammers were generally used in hand-forging;

(a) the hand hammer used by the smith himself, and
(b) the sledge hammer used by the striker, the smith's assistant.

For roughly forming the metal, direct blows from the hammer alone were used but for cutting off, forming and other finishing operations, various hand tools were necessary.

Chisels: These were used for cutting metals. They were hot or cold depending on whether the metal which was to be cut was hot or cold, and the main difference between the two was in the edge. The cold chisel had its edge hardened and tempered with an angle of about 60° while the angle of the hot chisel was 30° and hardening was not necessary since in any case the hot metal re-softened.

Tongs: The blacksmiths required a good selection of tongs with which to hold his work during various operations.

Process of Manufacture

In the fabrication of diverse articles, the ironsmith did not necessarily devise shapes and sizes according to the pattern and specification given or ordered by the customer.

The different stages of production of iron articles commonly met with are described below:

Shearing: The first stage of manufacture commenced with the shearing of iron or steel into desired sizes. This operation was performed with the help of a chisel and hammer. Before being sheared the scrap iron is to be heated into a red-hot condition.

Forging and Shaping: The iron pieces sheared were then kept into the burning furnace so as to make the scrap pieces malleable through heat making it easier for being hammered into the desired shape. As soon as the iron pieces became red-hot, it was picked up with the help of a tong and placed on the surface of the anvil and then hammered to form into a particular shape. This process of forging was repeated until the scrap piece took the final shape.

Tempering: The next stage was to temper the articles in order to impart a certain amount of strength to the tools and the implements to be manufactured. Only the skilled workers could do the job of tempering. Sharp-edged weapons or other objects required to be tempered for their appropriate use. Weapons and tools made of iron got rusty and blunt cr broke with use, if untempered. Tempering was done to prevent rusting, impart strength and sustain the sharpness of implements. Implements meant for woodwork and similar materials were subjacted to mild temper, while those meant for working on stone or metal were given a hard temper. The object to be tempered was firstly heated in a furnace to a pinkish red and then dipped into a solution of oil or water.

Gun Barrel Manufacture

India has a long tradition of gun manufacture and it was during this period that excellence in gun barrels took place. Special type of barrels damascened like the sword blades were manufactured.

In some parts of India the ironsmiths preferred for the material of their barrels the iron of old sugar-boiders. But where this was not available, they used the spongy, cavernous, and crude mass, which was first reduced from the ore. The fuel used was wood charcoal only. The forge was constructed exclusively for the barrel manufacture and was usually established at the gate of an oblong chamber, raised within the gunsmith's shop. The platform of this chamber was lower than the general level of the floor of the shop, by a few inches only at the gate, but it deepened by a regular slope to the farther end, making about a foot in difference between the front and the back. The sides and the end were closed by a straight, upright walling, and the top by a vaulted roof, pierced by a vent-hole for the smoke, which was diffused through the upper part of the shop and escaped by the roof and windows. The roof was generally horizontal.

The few-hole or space for the nozzle of the bellows, was invariably in the left-hand wall, on a level with the platform of the general floor of the shop, and about six inches from the mouth of the forge, which was open from the floor of the roof. The whole was built of brick, plastered both within and without clay mortar. There was no metallic back, and the tew-hole was in general metallic only in part, consisting of a narrow slip of iron, about ten inches in length, worked at one end into a broad longue, of which the edges were bent upwards, turned over, and formed the circle, or end of a tube, the rest being made merely of clay; and the narrow end of the slip running through the wall, and being worked into it, kept it, whole in place. The circle of iron was covered with a coating of clay, and then opening had a direction somewhat more downwards than horizontal, so that the stream of air from the bellows struck more directly upon the metal exposed to its action than that of European forge. The bellows were double and each pair consisting of a goat skin untanned but made pliable by being rubbed between the hands, and the hair side was left outwards.

For this appropriation the body of the animal was extracted by an incision, confined from the middle of the hinder part of the one thigh to the same point in the other.

A cylindrical tube of iron, or wood, was inserted into and tied tight in the neck, as the nozzle of the bellows. But the edges of the incision, by which, as before observed, the body was extracted, are stiffened, and kept apart by a long piece of wood sewn to each side, and these were connected with each other by thongs of white leather. These thongs, in the form of loops admitted the fingers and thump of the operator, who, with each hand, worked each pair of bellows inserted into the tew-hole by their nozzles, side by side, by opening and closing the mouth of each alternately, and pressing the air contained in t' eir cavities forwards and downwards. By this simple apparatus the air was blown in a continued stream, and a welding heat was soon reached. The whole forge was a parallelogram of such a length, as was nearly capable of receiving half the whole length of the longest harrel made and the bottom of the platform was covered with light ashes. A brick laid on its edge from the side wall just beyond the tew-hole projected into, and partly across the forge, and was met by another at right angles, placed in the length, from which resulted a space about 8 inches in length and 4 inches in breadth, bounded on one side by the wall, by the longitudinal brick on the other, and open to the mouth of the forge, having the tew-hole in the middle on the wall side.

In this place was placed the charcoal, of which the consumption was small and very little heat was lost through the fire being thus confined by the bricks, and by the layer of ashes, whilst the gun smith was capable of limiting the action of the heat to as small a surface of the work as he may desire.

The iron employed came from the smelting furnace, after, whilst hot received a few blows, which condensed. The first process consisted in cutting down the lump of iron, when heated, into narrow strips with a cold chisel. In this operation the iron lost one fourth of its gross weight, from which some idea may be formed of the large proportion of unmetallic matter which it contained. Each of these strips, separately, was brought to a welding heat, and worked smartly under the hammer of two men, on a anvil, when the slag was expelled by this operation, each strip was drawn out under the hammer into a strap about two feet in length, an inch and a fifth in breadth, and one fifth of an inch in thickness. One of these straps has its ends as brought together as to form a parallelogram, generally about five and a half inches long, and sufficiently broad to contain twenty lengths of the other straps cut for the purpose. Some of these were wedged in upon their edge, their faces standing parallel to each other, and to the long sides of the belting strap, of which the ends were previously welded together. The pie is then turned to its opposite face, and other pieces of strap are driven between the lengths, which were inserted on the opposite side, so as to make the whole light, and this object was further aided by its being put into the fire, lightly heated, and given a few blows upon both faces as well as upon the edges. It was next smeared over with a paste of clay and water, placed near the fire, and when dried, a light welding heat being taken at it. The whole was struck up into a tolerably compact mass by smart, but rather light hammering. It was later exposed to a stronger heat, the mass was vigorously and quickly hammered, and beaten out into four-sided bars, about a foot long and finger's thickness. These were each heated again separately, and drawn out into square rods, about a quarter of an inch broad on each face. One of these rods, neatly squared and free from scale was pushed into the forge fire so that the loose end penetrated under the cross fire bricks into the bed of ashes, and only the middle was exposed to the fire. When it got a red heat nearly verging upon white, it was withdrawn, and the loose end was quickly engaged in a square staple projecting from the surface of a 10 g of wood close to the mouth of the forge, so as to be held fast, the fireman having previously secured the opposite end between the jaws of a small pair of tongs with a slid on the stems, began to twist the rod on its axis from right to left, or in the contrary direction. The hotest part, or that which was fit for receiving a close twist was ordinarily two to three inches in length; and an assistant sitting close to the rod with a small pot containing cold water, and having a spout projecting from its side, poured a stream upon the part between the hot and the black portion of the bar at one end, and then quickly at the other, prevented the portion from receiving an imperfect twist; and by the same means the twisted parts were cooled. By one complete revolution of the rod, four winds were generated from the four angles of the square, and by the repetition of the process, through a succession of heat, the whole rod was then converted into a fine screw. Each heat furnished about two or three inches of twist, according to the ability of the gun smiths; and a considerable alertness, both in eye and in hand was required on the part of the worker to prevent the occurrence of twist in a portion of the rod not possessing a sufficient degree of heat to admit of it twisting kindly. Barrels wase called plain, pechdar, [or simple twisted, or Jouhardar or damasked liner intended to be produced, and which were distinguished by names taken from the place in which each variety was most affected, or from the nature, or figure, in which the lines were disposed—an Iran or Persian belonging to the former class, and Pigeon's Eye, Lover's Knot, chain and appertaining to the later.

To make an Irani barrel six or eight rods were required. In regard to the latter number four were employed of those twisted from right to left and the same number of such which had received the opposite twist, or from left to right. Every rod, after having been slightly heated; was lightly hammered on its two opposite sides equally: thus a rod had four distinct portion in its circumference of which two had the threads beaten down, and were somewhat flattered, whilst the two others had the threads standing, and retained their original roundness. The rods were then lengthened by being welded end to end and care was taken that each rod was made up of lengths of the same direction of twists with a little disturbance to the threads at the joints as practicable. They were next laid parallel to each other, their flat sides being in contact: rods of opposite twist were being disposed in alternate succession; as first one was twisted from left to right, next one twisted from right to left and so on. After this the extremities of the rods were brought into close contact, and welded together, each end of the general bar presented a wedge of about an inch and a half in length, flat on the upper and lower surfaces and narrow on the sides of edges. The middle rod forming the point of the wedge. In the whole extent of the bar the rods remained merely in contact the welding being confined to its extremities alone. The band or skelp was now ready for being formed into a hollow cylinder by being bent on twisted in a spiral line upon itself. This process began at the thicker end and continued to the thinner or muzzle, the lower extremity, or that intended for the breaching, being struck vigorously and perpendicularly down upon the anvil between the hammerings of every twist, for the purpose of jumping up the edges, and bringing them into close contact. When the twisting was so far completed that the edges of all the twists stood even, and touched each other, and the cylinder was nearly equal throughout its length, it was coated with thin paste of clay and water and was then ready for being welded. A welding heat was first taken in the middle of the cylinder and the edges of the twists were then heated were brought into intimate contact by being jumped up. After this the cylinder was returned to the fire and when the heat was well on, the twists that was jumped up was vigorously hammered, so that one round was ordinarily welded at a heat. At the commencement of the second welding heat, the jumping up was repeated, followed by the welding when the heat had been regained, and so on successively until the barrel had been welded up to the nozzle. The welding was recommended from the middle of the barrel and with a jumping intermediate between every two weldings, the process was continued to the breaching. This was repeated by commencing at the middle and welding to the breeching and after wards proceeded from the middle to the muzzle, during which an iron rod was introduced at each end and was used as a mandril; a third heat, but merely red, was taken at the whole surface of the barrel, which was vigorously hammered and was rendered regular and level. The barrel was then fixed horizontally through a hole in an upright post, and bored, after which its surface was filled, polished and prepared for bringing out the brilliant damasked lines.

9

Location of Some of the Copper Mines

On the basis of field Investigation and Geological records a list of regions has been drawn where copper was worked and sources open to smelters during this period in Bihar, Rajasthan and Uttar Pradesh:

1. *Surda, Singhbhum Distt., Bihar*: A large number of workings in the area indicates that it must have been an important mining centre for copper in old days. Ancient workings in this area traceable over a length of 600 metres. Gossan Capings seen, mineralization zone of quartz-chlorite muscovite schist.
2. *Laukesera, Singhbhum Distt., Bihar*: A persistent line of old workings.
3. *Mosabani, Singhbhum Distt., Bihar*: Old workings in this area are among the largest met within the belt and stretch between Mosabani and Badia. Veins are mainly formed of chalcopyrite. Copper ores have high nickel content.
4. *Dhobani, Singhbhum Distt., Bihar*: A number of large old workings, some of these measure up to 30 metres in width and upto 12 metres deep.
5. *Rakha Mines, Singhbhum Distt., Bihar*: A line of old copper workings extending for about 122 metres is seen in the south western flank of Rakha ridge. Mineralization is in the quartz-chlorite rocks. Old workings extend down to more than 50 metres. Some chalcopyrite, pyrite, chalcocite, along with some oxidized ore and native copper were met at about 48 metres down.

6. *Roam, Singhbhum Distt., Bihar*: Some old workings, encrustations of malachite seen in these old workings. Chalcopyrite vein here assayed 11.69 per cent Copper.
7. *Kendadih, Singhbhum Distt., Bihar*: A number of ancient small workings, with malachite stainings chalcopyrite and pyrohotite are seen in chlorite-quartz schist.
8. *Puranapani, Sighbhum Distt., Bihar*: A few old workings.
9. *Ramchandra Pahar, Singhbhum Distt., Bihar*: Old workings in chloritic schists, the main mineralized zone is 15 m. wide. Recent Geochemical Prospecting in the area on close grid pattern has shown high anomally zones along all the four lines of old workings.
10. *Rajdah, Singhbhum Distt., Bihar*. No. of old workings for Cu, a few of them fairly deep. Detailed mapping by GSI has indicated that the chlorite-schist is mineralized in a zone which is about 3 to 4.5 m. wide and extends for about 800 to 1000 metres.
11. *Baraganda, Hazaribagh Distt., Bihar*: Extensive old workings extending over a length of 1200 metres. Ancient mining was contemporaneous with that in Singhbhum copper belt. Average width 4 m., grade 1.5% Cu, Number of slag heaps found. Cu in slag 0.08 to 13.04%.
12. *Geyjadih, Hazaribagh Distt., Bihar*: Four mineralized zones in the area. Mineralization of the same nature as in Baraganda Mines. Old workings are found in 3 out of 4 mineralized zones.
13. *Agar, Alwar Distt., Rajasthan*: There are about three dozen circular lined shafts 10 to 35 metres in depth made by ancient people. Slag heaps occurring nearby, are estimated to be of the order of about 1 million ton and are found to contain up to 4000 p.p.m. of copper.
14. *Birat, Alwar Distt., Rajasthan*: A series of old shafts and pits extending for about 800 metres in strike are seen located along a brecciated zone. The old workings are accessible upto a depth of 35 metres.
15. *Baldeogarh, Alwar Distt., Rajasthan*: There are a number of old workings which are upto 10 metres deep and extend over a length of about 300 metres. At the

northern end of the prospect there is a large copper slag heap. Recent geochemical prospecting by GSI indicated that the mineralized zone extends over a length of 400 metres.

16. *Bhagoni, Alwar Distt., Rajasthan*: There are a number of old water-logged shafts more than 10 metres deep. The area is littered with slag heaps and mine debris. The area is under detailed investigation by GSI.

17. *Dariba, Alwar Distt., Rajasthan*: About 0.5 million tons of a rich grade copper ore has been proved by I.B.M. There are a number of old mines. Some are worked upto 75 metres deep and extend for about 400 metres. Mineralization consists of chalcopyrite, pyrrhotite, pyrite, appreciable quantities of Ni and Co indicated.

18. *Jodhawas, Alwar Distt., Rajasthan*: There are a number of old deep and broad inclines, shafts and trenches. The area was prospected by IBM with the little amount of work done, favourable results could not be obtained.

19. *Nalladeshwar, Alwar Distt., Rajasthan*: There are a large number of old copper workings which include a number of narrow, deep and well lines shafts also. Considerable amount of slag is found. Some samples from these show 2000 p.p.m. of copper.

20. *Mawaika Poora, Bharatpur Distt., Rajasthan*: Number of trenches about 3.5 metres to 9 m. wide and upto 12 metres in depth are seen for about 400 metres. Slag heaps found in the nearby area suggest that the smelting operations were carried out in ancient period.

21. *Bidasar, Bikaner Distt., Rajasthan*: A nine metres thick zone of pyrite bearing schist, assaying 1% copper, is reported to have been intersected in a bore hole near Bidasar. A stray nodule of ore collected from a place about 5 km. away from the village assayed 12% copper. Several slag heaps are also found in the area.

22. *Baleshwar, Jaipur Distt., Rajasthan*: There are about a dozen narrow, deep circular shafts which are located along a band of hard, black quartzose schist near the junction of Alwar and Ajabgar Series. Large slag heaps and occasional malachite stained rock fragments are

met within the area.

23. *Bagor, Jhunjhunu Distt., Rajasthan*: There are a number of old workings and old pit which is now completely filled up. Appears to have been the central shaft of a once prosperous mine.

24. *Ajitsagar, Jhunjhunu Distt., Rajasthan*: Several old pits occur about 1 km. south of Ajitsagar reservoir. The accessible portions of these pits show ferruginous and malachite stains. This place is reported to have been one of the largest mining centres in the past.

25. *Akwali, Jhunjhunu Distt., Rajasthan*: There are a large number of big old mines in the area. A strike length of 4500 metres is under detailed investigation by the G S.I. About 27 bore holes have been completed and under ground mining has been started recently. The area is potentially quite rich and there are fairly good chances of getting a sizeable reserve of copper ore sufficient to open a mine workable economically.

26. *Gurha, Jhunjhunu Distt., Rajasthan*: There are a number of old shafts and pits upto over 30 metres deep, found along a fault zone in the debris near these workings. Sulphides are seen disseminated in the quartzite. A gossan zone about 8 to 20 metres wide has been traced along about 800 metres, beyond which it is covered with dune sand. It appears that the mineralised shear zone continues south-west ward upto Girwari (27° 45': 75° 31') and even beyond it. At Girwari a number of old workings are located in phyllitic quartzites. Preliminary investigation is being conducted by G.S.I.

27. *Khetri Copper Belt between Singhana and Baba Jhunjhunu Distt., Rajasthan*: Extensive old copper workings have been known to occur in this belt which is about 20 km. long. Mineralization consists of chalcopyrite, pyrrhotite and pyrite. Traces of nickel are reported. Presence of very large slag heaps near Singhana suggests this place as an old smelting centre.

28. *Dhanaota, Jhunjhunu Distt., Rajasthan*: Indications of heavy mineralization confined mainly to phyllites, have been noted within the 'Baun Kui' (52 wells) hill near Dhanaota. A number of old shafts upto 35 metres in

depth are also found in the area. It is reported that mine was in operation till the end of 19th century. There are large heaps of slags in the hill. Mineralization is traceable over a length of about 750 metres, the width varying 100 to 250 metres. A thick gossan made up of limonite, goethite and pyrolusite is seen all along the northern flank of the hill.

29. *Nothoka, Sikar Distt., Rajasthan*: There are a number of old shallow shafts and prospecting pits. The mineralization is found along a north-south trending fault zone in phyllites. The mineralized [zone extends over a strike length of about 300 metres. Old adit about 30 metres in length has been stopped short of intersection with the mineralised zone. A sample of phyllites collected from here by Fox has assayed 0.6 per cent copper.

30. *Chanpura near Rakhabdeo Udaipur Distt., Rajasthan*: Extensive old workings for copper have been reported to be present along the hill tops.

31. *Dariba, Udaipur Distt., Rajasthan*: Several water-logged ancient workings are found near Dariba. A multi-metal load with about 10 per cent total metal content has been struck in a bore hole drilled by GSI. The different metals are copper, lead and zinc. C^{11} date of organic sample from this mine places the mining activity to least 360 B.C.

32. *Bafila Almora Distt., Uttar Pradesh*: An old mine exists below Bafila village. Loose boulders of vein quartz and talc bearing chalcopyrite, malachite, galena and pyrite are seen here. Copper slage is found all around the village.

33. *Dewaldhar, Almora Disit., Uttar Pradesh*: The area is reputed for its copper industry in the past. Recent detailed investigations have, however, failed to locate any lode of commercial value.

34. *Ganai, Almora Distt., Uttar Pradesh*: A number of old copper workings exist. Large scale mapping and detailed exploration by GST has not been successful in locating any lodes of economic value.

35. *Thela, Almora Distt., Uttar Pradesh*: A number of old

copper workings are present. Chalcopyrite disseminations in vein quartz and malachite encrustations are noticed and dolomite debris in the dumps around the old workings.

36. *Chamethi, Chamoli Distt., Uttar Pradesh*: There are three old adits. One of the adits, driven by Wilkon in the area between 1905 and 1910, is reported to have yielded fairly rich ore.

37. *Dandhakhan, Chamoli Distt., Uttar Pradesh*: Several old workings are found. Slag heaps are also found near the village and a sample from here assayed 1.65% copper.

38. *Dhanpur, Chamoli Distt., Uttar Pradesh*: Copper area have been extensively mined in the past. There are several old adits in the area in which small stringers and pockets containing lead ore, pyrite and chalcopyrite have been met with.

39. *Dobri, Chamoli Distt., Uttar Pradesh*: There are several ancient workings but only two adits have been negotiable. A groove sample from one of the adits assayed 0.68% copper and 1.61% lead.

40. *Pokhri, Chamoli Distt., Uttar Pradesh*: Pokhri is one of the most reputed localities for copper mines in Gharwal Himalayas. There are four old workings in this area, the most prosperous mine, being the Raja's mine. A peak production of 300 kg. ore per day bearing as much as 25% copper is said to have been reached at this mine in ancient times. Detailed investigation by GSI is progress.

41. *Asena, Tehri Gharwal Distt., Uttar Pradesh*: There are two shallow inclines. Occasional specks of chalcopyrite and pyrite can be seen along the incline. There samples from the area assayed 20% to 2.66% copper.

42. *Dharkot, Tehri Gharwal, Distt., Uttar Pradesh*: Some old workings exist in the hills. A few specimens collected from this locality indicated 0.45 to 0.98% copper.

10

Copper Metallurgy

Bronze, or copper with its alloys, is a landmark in the colossal strides of man to man to civilisation. Among the earliest copper objects in the old world, known today is a bead of native copper, dated to Ca. 6500 B.C. (Ca. 9000 years old) from Tape Ali Kosh (West Asia). During these 9000 years, metallurgy has advanced step by step by fits and starts, conditioned by the mineral resources of the area and the technological innovations of the people.

Extensive studies on the birth of copper metallurgy in old world and India have been made by Forbes, Coghlan Aitchison, Bhardwaj and Agarwal.

It is generally believed that Tal-i-Iblis near the copper rich Kerman range in Iran was one of the earliest known centres of copper metallurgy (Ca. 5000-4000 B.C.) and to date earliest evidence of copper metallurgy in the Indian sub-continent comes (Ca. 3300 B.C.) from Mundigak in Afghanistan and a number of places in Baluchistan, by Ca. 2600 B.C. There is also evidence of copper metallurgy from Pre-Harappan levels in Sindh and Punjab. There is a great spurt in copper metallurgy during the Harappan period (2300-1750 B.C.) in various parts of India. After demise of Harappans, copper metallurgy undergoes retrogression, as represented by chalcolithic cultures (Ca. 2000-1100 B.C). There is again greater activity in copper metallurgy as represented by Copper Hoard Cultures (1100-800 B.C.) and the other archaeological cultures also show metallurgical progress viz. Pandu Rajar-Dhibi and Mahishdal (Ca. 1000 B.C.) in Bengal, Sonepur and Chirand (Bihar) and Atranjikhera, Hastinapur, Kauṣambi and Rajghat (U.P.).

Copper Metallurgy

There is evidence of large scale mining and smelting during this period from a large number of copper mines in Rajasthan, M.P., Mysore, Andhra Pradesh and Bihar. At most of the places, where copper ores occur, one observes old pits, heaps of metallurgical slag, scattered pieces of malachite ores, and sometimes smelting crucibles, thereby suggesting that the copper ores were extracted and processed since long in many parts of India.

The author visited some of the ancient mining and smelting sites in Rajasthan (Khetri, Kolihan and Singhana), and collected ores for laboratory studies. The chalcopyrrite copper ore deposited are associated with ancient mining and metal smelting at Khetri, Singhana, Bairat and Babai in the Aravali region.

Technology means a scientific study of the technical principles, involved in making, working and shaping antique objects made of copper alloys, for lack of which, the ancient craftsmen could at least achieve limited results. Such studies make or could make enormous contributions to our understanding of the past.

For convenience, archaeologists and art-historians have described all types of metal objects made of copper and copper alloys as "bronzes". This is a misnomer, and hence a clear definition is necessary. Copper is one of the most widely used metals of antiquity. Pure copper is a relatively soft metal, and as a result, it was never particularly effective as a meterial for making objects of utility and art. It can, however, be considerably hardened, by mixing with other metals, such as tin, lead, zinc etc. When copper is alloyed with tin, it is known as bronze; and when copper is alloyed with zinc, it is known as brass. These alloys are not simple mixtures; but complex crytalline structures in which quite small variations in the relative proportions of the ingredients can radically alter the properties of the material. When bronze alloy contains lead more than tin content, the alloy is called lead-bronze. Similarly, when zinc is present in bronze alloy in significant quantity, or in quantity more than tin content, the alloy is known as zinc-bronze. Before the discovery of bronze, and later brass, copper was, of course, used for the manufacture of works of art and utility.

Metallic art objects are, in fact, products of the artistic and technological knowledge of the periods in which they were made; and a critical study of such objects would furnish a great deal of artistic, scientific and technical information concerning those times. Various techniques are involved in the creation of an object of art, and the artist makes use of the best available materials suitable for his purpose. These techniques and materials are no less important than the artistic expression itself; because, the evolution of such ideas and expressions are closely related to the materials and techniques adopted. It is therefore necessary to know as far as it is possible, the technical as well as material aspects of art or craft. Scientific Investigation based on chemical and metallographical studies may reveal some interesting facts about these objects hitherto unknown. The data so obtained may prove very useful in making their identification more precise; their interpretations more meaningful. It may even provide a valuable clue as to the date and provenance of a doubtful object. Such studies have yielded good results and enabled Indologists, Archaeologists and Museologists in describing metallic objects correctly. The importance of such data for the purpose of Indological and Archaeological studies and research cannot be underestimated.

Metallographic Detection

Metallographical method of investigation has found wide application in solving many problems related to technical studies of Metallic art objects made of copper alloys. "The structure of a piece of cast metal is affected by the way it cools from a melt; and so preserves of a record of the details of its history". Thus metallographic examination of the internal structure of an ancient metal reveals much about its technical history. Apart from the manner of casting, metallographic examination also reveals the manner of working after casting, the kind of heat treatment given in antiquity, and the nature of patination, whether artificial or natural. The fact that metals are crystalline substances forms the basis of metallography. As a pure molten metal, say copper, cools, small crystals begins to form from nuclei throughout the liquid and each crystal eventually will become a polyhedral crystal with its surfaces in contact with those of its neighbouring crystals; and these crystals, which are irregular in boundary, are usually referred to as 'grains'.

When a pure metal cools very slowly, the grains will be of approximately the same size, and roughly equiaxial. If however, cooling is rapid or uneven, as in the case of a 'chill cast' metal, it will be found that the grains near the surface are not equiaxial, but elongated. When a metal is worked lightly, by hammering whole volumes of lattice, it may become distorted temporarily, and then return to their original position after each blow; but when heavily worked, these volumes of lattice may become permanently deformed, and as a result of this, the crystal grain takes a new shape. This kind of distortion is known as 'slipping'. Under the microscope, 'slipping' will appear on the etched metal sample as fine parallel lines across the grains of the metal. Apart from slipping under severe hammering, the grains themselves will become flattened.

A cast or deformed metal, if annealed above the recrystallization temperature, may form a new grain structure. The new structure will have a smaller number of grain boundaries, and the metal will be soft, and more ductile than before annealing.

When a molten alloy of two or more metals cools, unlike pure metals, entirely different things may happen. "In many cases, instead of two metals being mutually soluble in one another, they will only dissolve in one another to a limited degree. As a result, two or more types of grain may be formed in an alloy, each being a solution of one metal in the other. These different solid solutions are referred to as 'phases' and usually occur, when metals of rather different atom size are alloyed. Bronzes and some brass are typical of this kind of alloy. In an alloy like cast-bronze containing a second element dendrites (or cast-crystals), which appear in the form of branches are commonly formed. In order to represent the possible changes completely, a metallurgist uses a constitution (phase) diagram, which is plotted in terms of the composition of the alloy, the temperature of the areas in which solid or liquid exist alone, or in mixture, and the regions where the different types of solid crystals form. "The diagram shows only the phases which exist, when an alloy has been held long enough at a given temperature to obtain equilibrium. A casting is made usually, too fast, to allow equilibrium to be reached, but the sequence of phases, nevertheless in general, follows the diagram". When a molten bronze containing about 15 per cent tin by weight is cooled, the first crystals of the solid phase will begin to form at about 960°C, and, this phase is called Alpha (α) phase, which will contain only about 5 per cent tin. More solids will be deposited on further cooling, and both solid and liquid will become progressively richer in tin. The last liquid, which occupies the interstices between the dendritic crystals, solidifies in the form of the β-phase at a temperature below 800°C. "With further cooling, β-changes to the closely related γ-phase, which in turn undergoes a profound change at 520°C, when it decomposes completely into an intimate duplex mixture of α and δ-called a eutectoid. The more tin a bronze contains upto about 25 per cent, the more eutectoid will be seen in its structure; beyond this δ-increase".

The size and regularity of the dendritic crystals vary greatly with the conditions of casting in a cast bronze. Large well formed dendritic structures are produced, when solidification of undistributed liquid metal takes place slowly.

The dendritic structure gradually disappears, as the copper rich dendrites absorb tin from the infilling when a cored bronze is annealed; and finally, a completely homogeneous structure of polyhedral grains is formed. If, on the other hand, a cored bronze is cold hammered or drawn, and left in this condition, the dendrites will be seen distorted, and would appear as isolated broken cast crystals. Such a bronze, if annealed, will result in the production of a fresh grain structure, finer than that resulting simply from the annealing of a cast and unworked bronze. "It is generally true that the more a bronze has been cold worked and annealed, and the finer will be grain structure. Additionally, 'twinned' grains will appear. These are pairs of identical grains, whose lattice structures are mirror images of each other formed on each side of a common plane".

The metallography of brass, in many ways, bears a close similarity with that of the bronzes. In brasses containing less than 36 per cent zinc, on cooling from the melt, as with bronze, coring takes place; so that the final appearance is one of copper rich dendrites in a zinc-rich infilling. A new phase known as he β-phas e is formed on cooling, when the proportion of zinc rises above 36 per cent; and the resulting material is thus one of copper-rich dendrities with a eutectoid infilling. Other changes in composition may take place, if the proportion of zinc is increased above 42 per cent; but brass of this nature, rarely occurs in antiquity. The results of the cold working and annealing of brass are similar to those described in the working of bronze.

"In practice, cast copper-silver alloys will be found to be cored. The effects of cold working of annealing on copper-silver alloys are similar to those on bronze". Although bronzes and brass objects frequently contain lead to form ternary alloys, the metal does not enter into solid solution with copper- and on cooling, lead may be seen distributed throughout the alloy as minute globules. Nickel, on the other hand, forms a solid solution with copper in all proportions; and the presence of nickel in small quantities cannot be detected in prepared sections". Arsenic also forms a solid solution with copper; but if oxygen is present, arsenic will form the compound, copper arsenite, which being insoluble in copper forms minute globules throughout the alloy".

Apart from these metals, Antimony and Bismuth also behave in a similar manner. The presence of oxygen and hydrogen in the molten metal had considerable effect on the soundness of a copper casting. Oxygen will react with copper forming cuprous oxide, and on cooling, it will appear as globules throughout the metal, and this will severely reduce the strength and its toughness. "To overcome this defect, copper was generally poled to reduce the cuprous oxide to metallic copper. If, however, poling was overdone, the molten metal would absorb hydrogen, from the furnace gases and at the moment of solidification, this would react with any cuprous oxide present to form steam, which trapped in the solid metal formed below holes, so weakening the casting —— ($Cu_2O+H_2=H_2O+2Cu$). The art of sound casting was to produc a compromise by correctly judged poling in which both the production of cuprous oxide on the one hand, and of steam on the other were kept to a minimum".

All the metallographical features stated in the foregoing pages can be examined through prepared sections of copper and various copper alloys under a metallographical miscroscope. The distinct and interesting features of the internal structure of these alloys can be recorded with the help of a suitable Camera for the work of interpretation. And these metallographical features will allow one to draw conclusions about its past treatment and techniques. Metallographical method of investigation of antiquities and works of art in metal are therefore very useful in revealing interesting information about the technique of their manufacture, which cannot be obtained otherwise.

Such a study was carried out with a special metallurgical microscope, which is necessary to examine adequately polished surfaces of metal by reflected light. The general procedure followed in this study for preparing a sample for metallographical study may be outlined below. A small clean corrosion free sample representing the entire core was cut from each object under study. Each cutting representative sample was extracted from the cleanest part (free from corrosion) of each object and each small sample thus obtained was then mounted on a block of plastic. A plastic block was made from 'specifix", which is an embedding material. "Specifix" is a two component acryl-plast consisting of a powder and a liquid. After mixing, the material sets in about thirty minutes at room temperature into a transparent, crack free substance of an appropriate hardness. "Specifix" is insoluble in alcohol, and diluted acids and bases; but readily soluble in acetone and carbon tetracholoride. This mounting of sample in a plastic block helps to handle it easily, while polishing, changing from coarse to fine grade to give an undistorted flat mirror polished surface. Polishing of each mounted sample was first carried out on carborundum abrasive paper specified to metal polishing from grade 220°C to 400°C to 600°C; it is essential to rub each time at 90°C to the direction of previous rubbing. This eliminates marks or stirations caused by abrasive paper. Too much pressure was avoided, since such pressure may generate heat, and bring about structural changes in the metal. It was then finally polished on a fine clean selvyt cloth fitted on a slow-running disc using 'Ap-paste' which contains an alumina of the highest quality, especially made for polishing samples for metallographic examinations. 'Ap-paste F' (fine) is especially useful for polishing non-ferrous and soft alloys, and the grain size is about 1/4 micron. Before examination, each polished sample was degreased by dipping it in a soap solution, and then washed under running water and finally rinsed with Absolute Alcohol, and then dried with hot air from a hair drier. Etching reagent suitable for copper, and almost all copper alloys are: 1. a 10 per cent aqueous solution of ammonium persulphate and 2. an acid aqueous solution of Ferric chloride (10 per cent of ferric chloride acidefied with 2 per cent hydrochloric acid). It is advisable to keep a stock solution containing Ferric choloride-10 gms., Conc. Hydrochloric acid-30 c.c. distilled water-120 c.c. and to dilute with distilled water as may be required. Over etching was avoided. Etching provides colouration, and easy detection of changes and crystal structure. Each sample was etched, either with acidified Ferric chloride solution, or with ammonium persulphate, but when satisfactory etching was not achieved by acid Ferric chloride solution, samples were etched with Ammonium persulphate solution, which produced good results, in some cases, particularly in the case of copper samples. Etching was necessary in order to reveal the inner metallic structures as affected by the differing rates of chemical attack depending upon the local composition, structure or orientation of the microcrystals.

Cire Perdue Casting

Cire perdue is a composite French term—Cire means wax and perdue means lost; translated into English the term would mean lost wax. It is used for all types of casting where wax is replaced by metal. Broadly there are two processes by which the replacement takes place. In one process a wax model is first prepared and then a mould is made around it. The mould with the wax model is heated, when the wax melts and drains off through an orifice provided for it. The space vacated by the wax is filled up with molten metal which is poured through the same orifice.

This process is known as solid casting process. In the other process, a model is made of clay or saw dust or similar other material and around it is modelled wax covering which is again covered up with clay or similar material. Here, melted metal replaces the wax covering and takes its shape. This process is known as hollow casting process.

History

Both hollow casting and solid casting processes belong to a very old craft tradition. In ancient literature, as early as Rig-Veda, there is a brief reference to hollow casting of images. In Yajur Veda, which is a later compilation, references are made to tin, lead and silver with which ornamental bowls were produced. Plenty of references are available in the Upanishadas also. In the second century B.C. Patanjali, in his Mahabhashya, an exposition of Panini's work, gives us a vivid picture of the artisans in the villages, the king's artisans as well as the gold smiths who manufactured gold and silver coins with punch marked designs and executed various metal ornaments in demand at that time. The Arthasastra of Kautilya and the Buddhist and Jain texts in Pali and Prakrit, give references of metals and metal smithy. Occasional references to the metal arts are found in Ramayana, while in Shantiparva of Mahabharata the duties of a king in relation to the artisans are mentioned and their guilds are enumerated. Manusamhita, Yajnavalkaya samhita and harivamsa also provide considerable information not only about the craft and craftsmen, but also about the mining operations and the artisans guilds. Although the above references provide much information relating to metal work and the craftsmen in a general way, they contain little information about the actual process of casting of metal figures. It is not until much after the Christian era that we come across detailed recorded instructions about the procedures to be followed in the matter of casting of images.

Agnipurana and Matsya Purna of the Gupta period have in some detail described the technique of casting bronze images and also about the selfless concentration with which the image maker must bring his work to completion. The most elaborate treatise on the lost wax process of metal casting is the Silpasastra (as treatise on image making) recorded in the 68th Chapter (22.23) of the Manasara believed to have been compiled in the Gupta period. In general, the metal casters of South India, follow the principles laid down in these Silpasastra. Tamil and Malayalam versions of them are being used today as technical manuals by the metal icon makers of Madras and Kerala.

In the Silpasastras are laid down elaborate and strict rules regarding the auspicious times when one can start the work the composition of the ingredients to be used for casting the various measurements of the different parts of the body and their relative proportions, details of the different stages of casting work, the various postures and inclinations of the figure etc. 'The rules consist of dhyanas of verses, describing the spiritual character and personality of the deity and the Lakshanas or formulae giving the measurements and suggesting the pattern and form of the figure in keeping with the dhyanas'.

Direct evidences of the antiquity of metal casting in India have also come to light and are now available in plenty. 'A small bronze statuette of a dancing girl found in the burried city of Mohenjadaro in Sindh shows that the process of lost wax (cire perdue) was known to India some 5,000 years ago and that it was already carried out with great skill and artistry. The bronze and copper pins unearthed at the sites of Chanhudaro, Mohenjadaro and Harappa of C. 2500-1500 B.C. were also probably produced by the lost wax solid casting process. Excavations at the ruins in Sirkap (Taxila) have brought to light gold and copper ornaments of the Hellenistic style which were produced in the lost wax metal casting method at about C. 327-325 B.C. The group of bronze images found at Kolhapur (believed to belong the 2nd century A.D.) and the four fragmentary bronze figures of Buddha excavated at Amaravati (believed to be of the 3rd century A.D.) are sufficient proofs to indicate that the Indian metal caster had achieved high technical skill and creative aesthetic superiority and perfection as early as the beginning of the Christian era. The small hollow-cast bronze bull from Sahri Bahol, Gandhara, of the 3rd -4th century A.D. is a notable example of India's metal icons of the early Christian era.

The 67½" high copper image of Buddha dated C. 2400 A.D. which was excavated in 1862 A.D. at Sultanganj on the Ganges is an outstanding specimen of the Indian metal craftsman's skill, imagination and sense of artistic perfection as well as a world land mark in the cire perdue hollow casting technique. The image is now preserved in the Art Museum of Birmingham U.K.

A number of Bronze images of the Buddha have been recently found in the ruins of a monastery at Sirpur in Madhya Pradesh, and they belong to the ninth-eleventh centuries A.D. The icon-workmanship of metal-smiths in Bengal had considerable influence on the metal working in Nepal and Tibet. In south India, during the Chola period (ninth to thirteenth centuries the cire perdue solid casting method was generally in vogue. Metal icons were fabricated according to canonically correct ornamentation and in attractive style, aesthetically erecting several figures even on one pedestal.

In this period there came out certain valuable literary works which give details of the icon-making methods. Among them the Manasollasa or Abhilasitarthacintamani attributed to the king Somesvara Bhulokamalla of the Chalukya Dynasty in the Karnataka area written in the twelfth century A.D. has given a very clear account of the icon-casting process as follows:

> 'According to the navatala measurement, as mentioned before, the expert should first prepare the image (i.e. the model) complete with all its limbs, yellowish in colour, beautiful to look at, and with the weapons and arms as prescribed.'

After placing the wax tubes of the length of a dhatura flower on the neck or the crown (of the image), (the artist) should besmear the image with refined clay.

To clay should be added charred husk, finely rubbed and cotton severed a hundred times and a little powdered salt. All these (when mixed with clay) should be ground on a smooth stone and (the paste) should be applied three times all over (the image).

The first layer (of clay) should be transparent (and thin) and dried in the shade. After a couple of days a second layer should again be applied.) when dry again, there should be the third coating thickly applied.

(One) should besmear the whole (image or model) with clay leaving the mouths of the tubes open; and the wise man should dry it (the clay coating) with care and judgement. The expert should first (i.e. before beginning the process just mentioned) measure the wax of the image which has to be made either in brass, or copper, or silver or gold.

Brass and copper should be taken ten times that of wax, silver twelve times, and gold sixteen.

(Then one) should encase the metal, either gold or one that is desired, with clay and coconut-shaped crucible (thus formed) should be dried in the aforesaid manner.

Next (one) should melt away the wax (from the mould) by heating (the mould) in fire and should afterwards heat the crucible in cinders.

Brass and copper melt surely with (the help of) cinders just kindled. Silver melts with (the help of) glowing cinders, while gold with (the help of) cinders flaming fivefold.

After making a hole with an iron, on the top of the crucible and holding it tightly with a pair of tongs, (one) should pour molten metal into the mouth of the tube in a continuous stream and stop when it is filled to the brim of the tube.

The adjacent fire should be put out for the purpose of cooling (the mould with the molten metal). When the image gets cooled naturally, the expert should break the clay very carefully.

Then the metal image varily resembles that in wax, endowed with similar limbs and other details.

The icons of the medieval period were indeed noted for their exquisite forms and lasting metallic lustre. The icon production, which was a flourishing metal-art particularly in south India, went hand in hand with the building of imposing temples in increasing numbers. The Vijayanagar empire was noted for brisk trade in a wide variety of bronze objects.

Cannon

The use of metal in weaponry has been known from the earliest times in India. Guns and cannons and the use of gun powder were unknown in India before the Mughal conquest. It is believed that guns were first employed by Babar in the first battle of Panipat in 1526 and were introduced in the Southern India by the Portuguese. A compilation of the 16th century Sanskrit work, entitled "SUKRANITI", deals with small guns, big cannons and the use of gun powder and fuse. In later years huge cannons were used in warfare in the country. One of the best examples is decorated "Fateh Rehbar" cannon of Aurangzeb at Golconda Fort. It is beautifully inscribed on the body and the bore of the cannon. Other massive cannon, one of the biggest in those days was the Malik-i-Maidan cannon which has been engineeered and built by a Turkish Engineer Mohd Bin Hasan Roomi in 1549. It is made of 5 metals in which copper is predominantly present and it weighs 55 tons.

11

Zinc Metallurgy

Metallic Zinc appeared much later in history than the other common metals. Copper, lead, tin and iron can be obtained as the molten metals by heating their oxide ores with charcoal (corbon), a process called reduction, in shaft furnaces, which were developed quite early in history—Zinc oxide, however, cannot be reduced by carbon until temperatures are reached well above the relatively low boiling point of the metal. (907°C) Thus the furnaces developed to smelt the other metals could not produce Zinc.

Archaeological explorations and excavations at Zawar, 48 km. South of Udaipur in Rajasthan have revealed that India was centuries ahead of Europe and China in the Industrial production of Zinc. Zinc smelting activity have begun before the first century BC and continued until the first quarter of the 19th century.

Producing Zinc of high purity marked an extraordinary feat. For Zinc poses problems in extraction as its boiling temperature is about 200 degrees C lower than the temperature at which Zinc oxide, the usual ore, is converted to Zinc and this Zinc vapour readily combines with oxygen in the air to form Zinc oxide.

The Zinc extraction process followed in India in antiquity anticipates by at least 1,700 years the salient features of the process, distillation per dexenum patented by William champion in Britain in the 18th century in terms of equipment design and composition of ingredients. This India can boast of being the first country to develop a sophisticated pyrotechnology for the production of Zinc.

Zinc was, and still is, extensively used in making brass, an alloy of copper and Zinc. Zinc in vapour form easily mixes with copper. Fusion of Zinc with copper increases the strength, hardness and toughness of the latter. When the alloy is composed of 10 to 18 per cent of Zinc, it has the pleasing golden yellow colour. It can also take very high polish and hence literally glitter like gold. On account of this, brass has been a popular alloy in our country and abroad for over 4,000 years.

Rasaratnasamucchaya, a 14th century manuscript mentions of Zinc as a metal. The principal mining being at Zawar. Travellers including Col Todd and the author himself had commented on the extensive heaps of debris that had accumulated at the site from past metallurgical activities there.

The old workings at Zawar is surrounded by dolomite hills from which sphalerite, a Zinc sulphide ore, was mined. These ore deposits are argentiferous, that is, they include small quantities of silver and are associated with galena, a lead sulphide. Along with Zinc, silver was also probably extracted in the past at Zawar.

The author has visited number of old workings around Zawar and has found that many of them were worked down to a depth of 120 m or more. Their deep shafts opened into many galleries and narrow tunnels following the natural configuration of sphalerite mineral veins at different levels. All the galleries were provided with $1\frac{1}{2}$ metre to 2 metre diameter holes for ventilation at regular intervals. Radiocarbon assay of the timber supports found at a depth of 100 m has given a time bracket ranging from 120 BC to 30 AD for the Zinc metal industry at Zawar. As these radiocarbon samples come from a depth of 100 m, it is possible to observe that the Zinc industry at Zawar may have started at least a few centuries earlier. More radiocarbon samples from Zawar are now being analysed at the research laboratory of the British Museum.

The heaps of debries remnants of the past Zinc smelting operations at Zawar are impressive. They are built up with spent Zinc distillation retorts, furance fragments and ash. A rough estimate of the debris at the site suggests that it could be of the order of a million tonnes. It appears that quite a large quantity of Zinc was distilled at Zawar over the centuries. There are a number of ruined houses at Zawar that were built of spent retorts. Spent retorts were carefully used as hollow bricks. These houses were probably tenanted by the metal workers of the past.

The excavations at Zawar brought to light a series of impressive, nearly intact, structural remains of furnaces that were used in the distillation of zinc in antiquity.

The furnaces consists of two parts consisting of a condensation chamber at the bottom and a furnace chamber at the top. The two chambers are separated by a perforated terracotta plate. The condensation chamber measures 65 cm by 65 cm and 20 cm in height. Here the zinc vapour is condensed into fine zinc metal powder. The perforated terracotta plate is a composite unit made up of four equal segments of 35 square cm each. It is four cm thick, study and well backed. Its perforations have two sizes: larger holes of four cm diameter each of which is surrounded by smaller holes of 2.5 cm diameter. Within the furnace the terracotta plate was found to be supported on a ledge on three sides in the furnace walls and a single solid terracotta pillar placed at the junction of four segments.

Up above the perforated plate was the furnace chamber. In it 36 charged retorts were arranged in six rows, six in each row, their condensation funnel tubes passing down into the condensation chamber through the larger holes in the terracotta plate. Thirty six zinc metal collecting vessels were placed one underneath each retort to collect and condense zinc vapour.

The furnace structures unearthed did not include their roof. However, curvature of the extant brick walls suggests that the furnaces probably had a domed roof with a chimney—like opening. This opening may have served two purposes: (1) it acted as an outlet for the flow of hot gases from the furnace and (2) it also facilitated continuous feeding of the furnace with fuel. The delicate nature of the thinwalled clay retorts suggests that the fuel used was light but efficient. It is, therefore, possible to conclude that only cow dung cakes were used as fuel. Cow dung cake fire can quickly reach a high temperature and also maintain it if there is a constant supply of fuel as well as oxygen. Oxygen is supplied to the furnace through an induced draught.

When the furnace is set fire to, hot gases formed within the furnace rise up and pass out through the top outlet inducing a natural draught of air. Fresh air goes into the furnace through the front opening in the condensation chamber and the smaller holes in the perforated terracotta plate separating the furnace chamber from the condensation chamber. Reduction of zinc oxide by carbon is a strongly endothermic reaction, with an absorption of a vast amount of heat (57,000 cal). A continuous supply of fuel and strong induced draught are, therefore, necessary to maintain the required high temperature of the furnace for extraction of zinc.

The heat generated in the furnace had to be efficiently transferred into the retorts. Therefore, the retorts had to be a small and thin-walled, made of a refractory material. When zinc was separated from zinc oxide it was in a vapour state. The vapour flowed down along with an excess of carbon monoxide, a highly reducing gas, into the metal collecting vessels placed in the condensation chamber. Here the vapour cooled down under carbon monoxide to form a fine powder of metallic zinc.

Retorts

The retorts used at Zawar were all small, cylindrical, 25 to 30 cm long, 10 to 15 cm in diameter, made by hand, of locally available river bank clay. The one end of the retort was tapered and pointed, the other futed to a coneshaped clay funnel whose condensation tube [measured 14 to 15 cm long. Before luting the clay funnel, the retorts were filled with the smelting charge.

The analysis of the spent charge reveals no evidence of sulphur in it. This is significant. The zinc ore at Zawar is a sphalerite, occurring as veins in an impure dolomite host rock. It now appears that prior to distillation, the ore was roasted to convert it to zinc oxide. The roasting required a temperature of the order of 900 to 950°C. At this temperature dolomite decomposes to calcium and magnesium oxides. The high content of these oxides is a clear indication that a high proportion of calcined dolomite was included in the smelting charge. Analysis of the spent charge further reveals the presence of potassium, aluminium, manganese, silicon, sodium and chlorine. Among these the first four elements are likely to have been derived from the ore and the associated gangue. However, the presence of over one per cent of soda and traces of chlorine in the spent charge is less easy to explain. It is not likely that they were also derived from the ore. It is more likely that common salt was added into the smelting charge. According to a 14th century text, Rasaratnasamuccaya, medieval Indian itarochemists specify common salt as one of the ingredients in the distillation of zinc.

The inclusion of salt in the smelting charge would serve zinc distillation in two ways. First, it would help to sinter dolomite and hence keep the charge within the retorts in a porous open texture. This would reduce the chances of zinc vapour getting blocked in any one part of the retort. Secondly, salt vapour would react with the silica and alumina of the clay in the retort wall to produce a thin film of glaze on its inner surface. This would prevent the loss of zinc vapour through the micropores in the thin clay retort wall. It is therefore no coincidence that salt was included in the smelting charge of the vertical retort zinc distillation process by Wiliam Champion.

Channelising Zinc Vapour

All spent retorts when freshly cut open show a clear cylindrical channel at the centre through the spent charge. This was provided to facilitate the free flow of zinc vapour down into the condenser vessel. It appears a cylindrical reed of 1.5 cm diameter was inserted into the retort after it was charged and the funnel part was luted on it. This reed helped to keep the charge within the retort when it was inverted and in the heat of the furnace the reed burnt, leaving behind a clear channel for the zinc vapour to flow freely out of the retort.

The analysis of the spent charge has not helped to identify the fuel that was used for reducing zinc oxide and also producing an excess of carbon monoxide. Rasaratnasamuccaya, however, mentions a number of carbonaceous ingredients used in the smelting charge for zinc distillation. They include turmeric, resins, soot, lac treacle, mustard and ghee (clarified butter).

Thus, continuous production of zinc at Zawar and probably many other sites in the Aravalli Hills in the Rajasthan area, rendered brass an inexpensive popular alloy in India. Even today the Indian village folk are justly proud of their rows of glittering brass vessels in their homes.

Recently a team of archaeologists of M.S. University, Baroda, officials of Udaipur and the research laboratory of the British Museum had studied the Zawar sites. The British Museum helped with investigations like radio carbon dating of wood samples recovered from mine shafts.

12

Gold and Silver

Gold

Gold!......Never was a metal destined to play so sinister a role in the ages-old history of mankind. For the sake of Gold blood-thirsty wars were raged, nations and states were annihilated and monstrous crimes committed. No words can adequately describe the sorrow and suffering caused by this beautiful yellow metal.

The history of gold is the history of civilization. The first grains of this metal fell into the hands of humans several thousands years ago, and almost immediately came to be considered precious. In ancient times Egypt was believed to be the richest country in gold. Excavations at the sites of former burial grounds support this view.

The first gold coins appeared approximately two and a half millennia ago. They were made in Lydia, a powerful slave-owning state in Western Asia Minor.

Silver

The ancient historian Herodotus writes that in the 5th century B.C. the Persian King Cyrus, when on the march, kept his water in "Sacred Silver Vessels". Indian religious books describe how water was purified by immersing white-hot silver in it. In many countries silver coins were traditionally thrown into wells during consecration ceremonies.

Perhaps the purifying action of silver can be considered the oldest occupation of this metal.

Silver and Gold in Literary Sources: Silver Rajat (Rajat-hiranyam, Arjun) and gold (Hiranya, Harit) are mentioned in vedic literature, showing that Aryans were acquainted both with silver and gold. The use of the term Rajat-hiranyam might be a pointer to electrum (natural alloy of gold and silver). Taittiriya and Kathaka samhitas and shatapatha Brahman also mentioned the use of gold and silver. Mentions of Suvarna, Shatman, Nishka and Hiranyapind are taken as some sort of coins in gold electrum or silver.

Arthashastra mentions use of silver and gold including their mining, smelting and minting of their coins. It also describes the features of good, bad, pure and impure silver and gold and mentions the duties of the state's gold smiths.

Greeks and Romans also give account of Indian silver and gold. Curtius, Strabo and Philostratus mention the use of these metals in Ancient India. The account of Old Testament shows that Ophir trade mentions silver and gold as items exported from India. Indika of Megasthanes also mentions gold and silver mines of India.

Archaeological Evidence of Use of Gold: Gold was among the first metals known to the Prehistoric man. Its native occurrence, glittering yellow colour, simplicity of treatment in obtaining it, high ductility and easy in shaping small ornaments without melting and annealing, encouraged early man to utilize it.

In India, the antiquity of this metal goes back to Harappan period. Gold ornaments are known from Lothal and Rajodi (Harappan and late Harappan, respectively). Tekkalakota (Neolithic-chalcolithic, Phase I). Diamabad (Chalcolithic Levels). Kausambi (S.P. II, ca. 300 B.C.). Megalithic burials of South India and later South Indian urn burials from Adichanallur. Gold foil beads and small gold objects are reported from Early Historic period from Nevasa, Nasik and Jorwe, Pandu-Rajar-Dhibi, Vaisali, Nagarjunikunda etc. Except the Harappan period, gold was a scarce metal all through.

In Deccan, gold was used for making ornaments at least by ca. 2000 B.C. Regular use of gold is evidenced from Kushana period, though evidence of gold mining goes back to Neolithic times.

Occurrence of Silver Ores in India: Silver is not reported to occur in India as a primary mineral. Sources of silver may be attributed to either of the following secondary sources:

(i) Silver associated with gold from Kolar gold mines in Mysore.

(ii) Silver associated with galena (argentiferous).

Silver associated with gold is produced in considerable quantity today at Kolar in Mysore and to a much smaller extent in Anantpur in Madras.

Silver is obtained from lead ores from various localities in India and its neighbourhood. In Bihar and Orissa it is reported from the districts of Bhagalpur, Manbhum, Monghyr and Singhbhum. Here silver has been found associated with lead but it does not occur in considerable quantities. Ancient mines of Zawar, 15 miles south of Udaipur in the Aravalli region of Rajasthan, are reputed to have been worked in ancient period to obtain silver from the zinc blende argentiferous galena. In north Burma, Bawdwin mines (shan states) of argentiferous galena are of special economic importance. It supplied large, quantities of silver and from the slag heaps around these mines, it looks that only silver was extracted and lead was left out. Though it is primarily a lead-zinc-copper sulphide ore, its name Bawdwin' (silver mine in Burmese language) indicates that main purpose of its mining was silver extraction. During 1951-52, lumps of galena (argentiferous) were met at Metri, some 25 miles north west of Ballery in Mysore, with good content of silver. However, it seems that except Bawdwin mines of Burma, the other deposits are not of perennial nature.

In addition, Faranjal mines of Afghanistan, some mines in Armenia and many lead mines of Persia have been pointed out by Marshall for obtaining silver in Ancient India Reference may also be made of important sources of silver and lead in Bacteria and transoxania.

Occurrence of Ores of Gold in India

Forms of Gold Occurrence: As 'placer' metal (alluvial or detrital gold) in the auriferous sands and gravels of a large number of rivers which pass over auriferous rocks. Rivers of Madhya Pradesh, Maharashtra, Bihar (specially the region of Singhbhum and Manbhum distt. in the valleys of Suvarnarekha river) deserve special mention. That alluvial gold washing was carried out in antiquity is substantiated from the fact that gold dust was paid by Indian provinces as an attribute to Darius the Achaemenid.

In Quartz Veins or Reefs: Important sources of gold in this form are the Kolar gold mines in Mysore which produce 99 per cent of India's gold output. In addition, Hutti gold mines 6° 12′:76° 43′) in Raichur district (Mysore) and Ramagiri area in Anantpur district and Wynaad area in Nilgiri district (both in Tamil Nadu) also deserve mention. The workings at Hutti contain the deepest ancient shafts known in India; one of them being 620 ft. deep. Old workings are also known in the province of Tamil Nadu (Madras), old workings in south east Wynaad are supposed to have supplied gold to many invaders of India.

Gold Metallurgy: In the beginning gold nuggets were collected from the surface and gold was mined by 'open cast' method. Later, the method used for mining appears to have been fire setting, where rock faces were heated and then cracked by throwing water on them. This helped in sinking deep shafts in quartz veins. Stray finds among old workings, two radio-carbon samples from Hutti and field investigations indicate that shafts through the quartz veins had reached considerable depth by the opening of Christian era. Over 600 ft. deep mines which were worked in ancient times have been recorded at frequent intervals along several miles of reefs of Kolar and Hutti gold mines.

Considering, that gold from Kolar and Hutti gold mines is of light colour (being electrum-native gold-silver alloy) and similar gold objects have been recovered from Harappan sites, it is justifiable to infer that surface collection of gold nuggets had started by Harappan period.

Methods of Extraction of Silver and Gold: *Silver*: From above, it is clear that important sources of silver could have been either argentiferous galena or electrum.

Production of Silver from Argentiferous Galena: Production of lead from lead ores is done in three stages (i) smelting of galena (ii) purification of crude lead and (iii) desilvering of the soft lead. Smelting is done by roasting process, when galena is decomposed and most of the sulphur is oxidized and some is left as lead sulphate. Some galena remains, but most of it is oxidized to lead oxide (litharge). When the temperature is raised litharge, lead sulphate and galena interact to form lead. Alternatively, galena is first roasted until practically all the galena has.

Heating silver with lead in skull which consists mainly of calcium phosphate, practically resembles cupellation process for the purification of silver. There are a number of other literary references to the extraction of silver in later Indian alchemical works.

On the consideration of the above methods, we may say that in ancient period lead ore was first smelted to crude lead and silver concentrated by a process somewhat similar to Pattinson's process. Crude silver was then cupelled. In antiquity the practice seems to be to smelt silver ores with lead to ensure complete extraction of all the silver.

Method of Production of Gold: Since gold occurs in metallic state, after mining, the gold is simply melted and cast in to requisite shapes. However, in mining, gold particles are separated by crushing and sifting the ore by making use of high density of gold. In between reef gold and placer gold, extraction of gold from the reef would have been more difficult. As mentioned earlier, there is some evidence of Kolar gold mines having been worked in remote antiquity (probably in Harappan period). In the beginning, the metallic nuggets might have been hand picked from the surface out crops and followed by breaking the rocks by fire-setting. The selected pieces were crushed to small size in mortars and hand querns. The pulverized ore was treated with water and gangue material and gold was separated (reef mining as conducted in Ancient India shows a depth of 620 feet).

In the case of alluvial deposits, the sand and gravel are agitated with water in pans, the rocky matter is floated off and gold particles are collected at the bottom, alternatively, the alluvial sand was passed through agitating cradles or sluices, provided with obstructions (to retain gold). In this process lighter sand is washed away and gold particles are retained.

Gold particles whether obtained from alluvial washings or separated from reef gold were fused into small ingots in clay crucibles over charcoal fire. For this, a temperature of 1063°C would have been necessary.

First copper and iron, ivory and agate, later Gold and Silver and gems became the medium of adornment. Apart from the primitive art of filgree the highly evolved art of enamelling also reached a high degree of perfection in this country. Traces of it are found even at taxile and the art reached its pinnacle during the Mughal period.

Gold and Silver were also used for making coins and other decorative items like flower vase etc.

Gold and Silver have been lavishly used for making jewellery for adornment. Gems and jewelry, ivory and gold, silver and semi-precious stones have been the pride of the East for ages. The name of India conjures up a land of colour, beauty and sparkle, with its charming countryside and the glittering jewelry that adorns the person of her women-folk.

The origin and significance of jewelry go back to the earliest times, and one sees it through the ages as a remarkable manifestation of aesthetic urge, the inventive genius, the technical skill, and the prosperity or decline of the nation. Its development, from the Chalcolithic period to modern times, is a story of the ingenuity and progress of man in his search for new and enduring materials, resulting from his desire to make himself, and especially his womenfolk, attractive. This desire was common to both men and women in early times as is evident from the fondness of both sexes for ornaments.

The sources that provide the material required for putting together the story of jewelry from ancient times to the present day are to be looked for in the valuable finds from archaeological excavations dating back to the hoary past, the hymns of the Vedas handed down, first by word of mouth from generation to generation and later recorded, the great epics—the Ramayana and the Mahabharata—the magnificent and enduring monuments of the visual and the plastic arts, and the great works of classical Sanskrit literature. The valuable material available from all these sources helps us to reconstruct the pattern on

Indian Life in ancient times, or in ancient civilization, as we know it today. That pattern, which is a combination of oral tradition, pictorial and plastic art, and literature of a very high order, presents a wonderful panorama of the resplendent civilization that has come down to us through the centuries and of which Indians are naturally proud because of the variety of its colour, form and content as well as its enduring quality.

Jewelry is constantly mentioned in the Rig-Veda. The Maruts decorate their persons with "various ornaments," "they are richly decorated with ornaments," and "shining necklaces are pendent on their breasts." The Aswins are adorned with "golden ornaments," and the Asuras likewise have "plenty of gold and jewels." The sage Kakshivat, the reputed author of several of the hymns of the Rig-Veda, prays for a son "decorated with golden earrings, and a jewelled necklace," and largesses of "gold and jewels," to the priests and Brahmans are constantly mentioned. In the Nirukta, or Etymological Glossary, forming one of the Vedangas, and in the grammar of Panini, who is supposed to have lived in the fourth century B.C. the names of various kinds of jewels are given, which are identical with those still in parlance throughout India. Manu minutely defines the nature and duties of the jeweller, and the fines he is to pay for piercing precious gems, such as rubies and diamonds, and for boring inferior gems improperly, and the punishment due to him for debasing gold. The references to jewelry in the Ramayana and Mahabharata are too numerous for quotation. Sita is represented as arrayed for her marriage with Rama in a light sari—like garment of a rosy red colour embroidered with gold, and with jewelled butterflies and other bright ornaments in her raven black hair. Her ears are resplendent with gems, she has bracelets and armlets on her arms and wrists, a golden zone binds her slender waist, and golden anklets her anklets. She has jewelled rings on her fingers, and golden bells on her toes, that tinkle as she walks with naked feet over the carpeted floor. In the Mahabharata, at the gambling match at Hastinapura, Yudhisthira is described as losing first "a very beautiful pearl, next a bag containing a thousand pieces of gold, next a piece of gold so pure that it was soft as wax, next a chariot set with jewels, and hung all round with golden bells, next 1,000 war elephants wigh golden howdahs set with diamonds, next 100,000 slaves all dressed in good garments, next 100,000 beautiful slave girls, adorned from head to foot with golden ornaments, next all the remainder of his goods, next all his cattle, and then his whole kingdom, excepting only the lands he had granted to the Brahmans." Sudraka, the royal author of the Hindu-drama of "The Toy Cart," and who lived in the first century B.C. or A.D., describes the jeweller's atelier attached to the house of a courtesan: "Where skilful artists were examining pearls, topazes, emeralds, sapphires, lapis-lazuli, coral and other jewels. Some set rubies in gold, some string gold beads in colored thread" [exactly as is done now], "some string pearls, some grind lapis-luzuli, some cut shells, and some turn and pierce coral."

The old vocabulary of Amara Sinha, one of the "nine gems" of the Court of Vicramaditya, B.C. 56-A.D., quoted by Rajendralala Mitra, gives a long list of names for crowns, crests, and tiaras for the head; of rings, flowers, and bosses for the ears, of necklaces of from one to one hundred rows of gems, of all shapes and patterns of armlets and bracelets; of zones and girdles for the waists of men and women, of anklets, and other ornaments for the legs, and of rings for the fingers, and bells for the toes.

In tracing the history of jewelry from the earliest times to our own, we find an unbroken chain of evidence to show that it has continued to exercise a fascinating influence on the minds of people, in spite of the variations of its forms or the materials used for making it. Many names and designs that were in vogue in ancient times are still in existence. Some names of ornaments of the modern age are ordinary words, while those in use previously were poetical or fanciful. This fact indicates that in the modern age of scientific advance there is a tendency for people to be prosaic and practical rather than poetic or imaginative.

Every article of jewelry referred to in the ancient texts had a meaningful name, and every design was a fulfilment of an artistic idea. The designs were of a varied nature, they were abstract, grotesque or fanciful, or with motifs delicately and elaborately wrought. Each period had its own conventions as witnessed by the varying patterns that prevailed from time to time.

Processes

Annealing: Gold and silver becomes hard by hammering or by drawing it often to make a wire out of it. In case the process is continued without re-heating the metal, it is likely to break. To prevent metals from becoming brittle re-heating is done. This process of softening metals is called annealing. By doing this, metals become ductile and easier to work on.

While annealing a metal sheet or wire of any gauge, sunars heat them to a dull red in between the soft flame of charcoal. While annealing a keen eye is needed so that the metal does not melt again. The air is supplied through the blow pipe and the metal is evenly heated. When long thin wires require annealing they are first coiled and then tied with an iron wire so that these do not stretch. This is placed in the charcoal flame. Since the wire is very thin there is every likelihood of its melting at any spot with an unbalanced flow of blast. But with deft training the goldsmith manages this easily. The next is judgement to know when to remove the metal from the forge after it has been evenly annealed. Heated metals are either left to cool or dipped in water, if the work is to be taken up soon. But artisans never dip the metal into water when it is red hot. If this is done the metal loses its softness to some extent. In case of large pieces metals are held with pincers on one side and the hammering is taken up. When the pieces are small the artisan holds one end of the mildly hot metal in one hand while he shapes and works on it with the other. It is not necessary that the metal should be hammered soon after it looses redness. The annealed metal can be worked any time. There are no chances of its breaking or cracking. After thirty to forty gentle strokes, re-annealing becomes necessary to avoid cracks, splits or stresses. In case of drawing wires, these can be drawn from draw plates for about ten to fifteen times at each annealing. At the same time it is the judgement and experience of a goldsmith, which in the end counts so much.

Designing: Annealed metals are shaped and designed step by step according to the ornament. It is clear from minute observations that the whole range of ornaments are mainly prepared either from a sheet or wire or pellets, or all together. While preparing ornaments several types of silver and gold sheets and wire of different gauges and dimensions are required. These are made by the goldsmith. Sometimes the sheets are cut into circles, hexagons or octagons, ractangles, semicircles, squares, strips, triangles, or any other geometrical shape which may strike the artisan or as may be required for an ornament. These shapes are again worked to form various depressions, convex and concave impressions, ornamental and floral designs, twining creepers, zig-zag engravings, and a number of other geometric appliquings. Animal heads particularly of elephants and lions and bird and fish figures may be carved on them. Bezel boxes are also cut at various places of an ornament where stones have to be set.

For designing these articles a number of implements are used. Metal and wooden moulds for carving of various depressions are required to give a round shape to the sheet. For large depressions bossing hammers and mallets are used. For making small depressions the dapping tool called 'padernoo' serves. Engravings are done with pointed chisels, scrapers and scribers. For the embossing of metals a number of iron and bronze dies are used. Sometimes simple dies of bronze are prepared by the goldsmiths.

Wires of different thicknesses in round or half round shapes are prepared and then plaited, knitted or twisted according to the design. Formation of these designs are considered easier. Thin wires are used for making chains, linear appliqued shapes and oval hook links. Thick wires are used to make strips by hammering and cutting them according to the design.

Soldering: Goldsmiths specialised in soldering work. After shaping different pieces of an ornament, these are deftly joined together. This is sometimes done at the initial or middle stages and sometimes when the skeleton of an ornament is ready. Joining is done piecemeal or at a stretch.

When soldering is done a specially prepared metal is placed in between the joints. The whole piece is evenly heated. This solder has a lower melting point as compared to the metal that has to be joined. When the solder fuses at a melting point it soon blends both the pieces and forms a firm joint.

Goldsmiths have no book knowledge about melting temperatures of pure gold and silver and their alloys. It is their subtle experience attained by regular and concentrated efforts. This is learned under the apprenticeship of an ingenious craftsman.

Gold and Silver in their pure forms melt at 960°C and 1063°C. As the base metals are added to them in gradual progression, the alloys become hard and their melting point decreases. Goldsmiths do not need to make an alloy of silver or gold. Goldsmiths have been preparing ornaments of the purest gold and silver. But at the same time they have the skill of preparing ornaments of metals containing impurity. They make silver ornaments which have one-fourth of base metal. Some can skilfully make ornaments out of silver even with half of the base metal.

Similarly gold can be used in various proportions of alloy. While making alloys, lustre of the metals is maintained though slightly different to those which are wrought from the purest or slightly impure metals. Preparation of a solder for making the joinings and the technique involved in achieving its easy flow and adherence to the base metal is said to be known to a very few skilled craftsmen.

Goldsmiths know that joining of ornament pieces, with solders having smaller quantity of base metals in them, form a hard and strong joint. As more base metals are added the joints become softer and break with a slight bend.

Solder for Silver Ornaments: Some alloys have nine parts or more of silver in one tola. These can be easily soldered. To carry out soldering on this common range of metal another alloy of silver would be required. This may have a lower melting point to the one being used for the ornament. Goldsmiths prepare these solders in different proportions of silver and zinc. For silver, the solder, commonly used is made by adding one part of zinc and four parts of silver. For making a solder metal is debased. The solder containing 3 mashas of zinc and 12 mashas of silver is the best. Other solders containing one part of zinc and 6 parts of silver are known and are used for superior silver. Solder prepared by adding 5 parts of zinc to 48 parts of silver is used for enamelling work.

Solder for Gold Ornaments: The solder commonly used for

gold soldering is prepared by adding eight parts of gold and one part of either silver or copper or both.

Preparation of Solders: Goldsmiths melt the gold or silver in a crucible. A small amount of borax is added. When the metal attains the molten state, other metals are added, in certain ratios. The metal is then mixed in the crucible with an iron rod. The molten alloy is poured back. The lump of alloy thus formed is flattened to form a thin sheet. Next the sheet is cut with snips, into very small oblong and square pieces. These are about one tenth of an inch or even smaller. These tiny fragments of solder are collected in a small container. This contains borax powder. A few drops of water are poured to form a thin paste of it. This is called to form a thin paste of it. This is called flux. A small piece of cotton held with tweezers is used in mixing up the bits of solder with flux. The ornament which is to be soldered, is also coated with flux. Then a scrap of solder is placed on or in between the joint with tweezers. The article is left to dry. If the ornament is a large one and needs soldering at more than one place, it is first given a close fit with an iron binding wire. The portion to be soldered is then smeared with flux. On the wet thin paste, pieces of solder are placed at the spacing of about 3/10th of an inch. The solder pieces will stick to the coated portion. Then the skeleton of an ornament is dried in the low heat or kept apart to dry.

Spacing generally depends upon the size of solder scraps and as well as on the thickness of the ornament. Some times joins are required to the depth of a quarter of an inch or more as we find in a typical bracelet. He decides the spacing according to the article and the size of tiny bits of solder available. More of solder would make the precious metals impure, and a little untidy. This may benefit the jeweller and bring him more profit. Particularly in gold work use.

Preparation of Solder for Joining Grains of Gold: A lump of gold alloy meant for solder is flattened into a very thin sheet. The sheet is made two-fold. Again delicate hammering continues. After sometime another fold will be given to make the leaf four-fold. To avoid sticking of folded portions water is sprinkled between the folds every now and then. This leaf is cut into a number of tiny scraps with snips. With utmost patience these are placed and stuck to the piece of work and further processes are taken up.

These tiny thin bits of solder melt at a very low heat. In case more heat is given tiny pellets, which need soldering, may themselves melt, and thus the whole work may be spoiled. Therefore special attention and vigilance is required. Soldering in grain work is not done by directing the flame on the work. It is carried on by burning specially prepared thin dry sticks of wood. These sticks are called bhalethi. These are very inflammable. The flame is brought close to the ornament. With the heat tiny bits of solder melt and stick to the grains. At the time, when soldering is in progress, even the slightest blow of air may cause damage. It is for this reason that sunars think their old type of forge more useful. Its walls stop the flow of natural air and protect the delicate piece of work from any harm.

Sunars say, it is a matter of great skill to join articles with the minimum of solder. To make use of a hard solder indicates great craftsmanship. But it is difficult for a beginner. There are artisans who claim that they can solder gold ornaments with gold of the same purity, though it takes minute labour. They feel it is not worthwhile to put in such labour during a days hard as these.

Preparation of Flux: Gold and silver in purest forms do not oxidize during heating. Because of the copper and other contents in them, there appears a thin layer of greyish and somewhat light reddish colour on the surface of metals, when they are heated. This layer is called an oxide scale. This oxide prevents solder from easily reaching the joints. To overcome this, pieces of the ornaments are dipped into a pickle. This is prepared by adding one part of sulphuric acid to 10 to 15 parts of water.

Soldering of pieces of an ornament cannot be taken up unless they are heated. But while heating an oxide scale will hinder the flow of solder. To prevent this, portion to be soldered is smeared with flux. On one hand it is done to avoid oxidizing of metals and on the other it promotes the flow of solder around the joints. A general formula for preparing flux is given here. Ammonium Chloride, borax and crystalline nitre are taken generally in ratios of 6:1:3. These are ground separately to make a thin powder. The powders are mixed properly in water to the consistency of a thin condensed paste. This paste can be used as flux. Some sunars take up further process to prepare flux and make sticks of it. This is done by boiling the paste in a tin bowl for about fifteen minutes. As soon as the mixture thickens and starts forming into bubbles the bowl is taken aside and left to cool. The paste becomes hard within a few minutes. This hard white substance is removed from the bowl and again ground. The powder prepared this way is called borax by sunars and used as flux. This stick can be rubbed on the clean slate by using drops of water and the thin paste produced this way is used in the same way as flux.

Soldering Process: The units of an ornament are cleaned well. They are tightly tied with an iron binding wire. The whole piece is properly coated with flux, having solder fragments firmly stuck to it, and finally dried well. The combined piece is ready to be heated and soldered. If the piece is wet, drying and soldering is done subsequently. Care is taken that it is warmed under a low heat. If heated abruptly it is likely that bits of solder may fall. Artisans make sure that the piece has dried up and the bits of solder are sticking at their original places. Then only the piece is brought to a soldering temperature. For this blow-pipes are used. This helps in throwing the heat at the right place. Goldsmith is careful that heat is not directed at the solder, otherwise, tiny bits of solder would melt and fall. Even if one part of the joint receives heat enough for soldering and the other does not, the solder adheres only to the hotter part and the join is not a success under no circumstances is the piece made red hot, as it may melt the metal. The smallest flaw in soldering, particularly of minute and elaborate pieces, ruins the whole work. In that case, the work is taken up afresh. Great deftness is shown in carrying out the soldering job as there is a very minute difference in the melting temperatures of the metal and solder. Whatever be the processes involved, it is the correct judgement of an artisan, who knows what to do and when.

Simple articles can be soldered by putting them in charcoal without giving any support. A single soldering of simple articles like rings can be done on a big piece of charcoal.

Methods Adopted to Prevent Disturbing of Earlier Solderings: Often, a piece of an ornament needs a number of solderings. In such pieces, earlier soldering would be disturbed when these are heated for further soldering. To overcome this, two methods are adopted by artisans. The suitability of either of the methods adopted by them depends on the work. When the soldered piece is dipped for few minutes in the acid mixture, joints become firm. In this way, the risk is covered.

Abul Fazl in Ain-i-Akbari describes in detail about some aspects of silver and gold metallurgy.

The Method of Separating the Silver from the Gold: "They melt the composition six times, three times with copper, and three times with sulphur, called in Hindi. Chachhiya. For every Tola of the alloy. They take a masha of copper, and two mashas, two surkhs of sulphur. First they melt it with copper, and then with sulphur. If the alloy be of 100 tolas weight, the 100 mashas of copper are employed as follows:

> They first melt fifty mashas with it, and then twice again twenty-five mashas. The sulphur is used in similar proportions. After reducing the mixtures of gold and silver to small bits, they mix with it fifty mashas of copper, and melt it in a crucible—They have near at hand a vessel full of cold water, on the surface of which is laid a broom like bundle of hay. Upon it they pour the melted, and prevent it, by stirring it with a stick, from forming into a mass. Then having again melted these bits, after mixing them with the remaining copper in a crucible, they set it to cool in the shade; and for every tola of this mixture two mashas and two surkhs of sulphur are used i.e. at the rate of one and one-half quarter ser (1-3/8 ser) per 100 tolas. When it has been three times melted in this manner, there appears on the surface a whitish kind of ash, which is silver. This

is taken off, and kept separate; and its process shall here after be explained. When the mixture of gold and silver has thus been subjected to three fives for the copper and three for the sulphur, the solid part left is the gold. In the language of the Panjals, this gold is called Kail, whilst about Dhite, it is termed pinjar. If the mixture contains much gold, it generally turns out to be of 6½ ban, but it is often only five and even four.

In order to refine this gold, one of the following methods must be used. Either they mix fifty tolas of this with 400 tolas of purer gold, and refine it by the Salone process; or else they use the *Aloni* process. For the latter they make a mixture of two parts of wild-cow dung, and one part of salt-petre. Having then cost the aforesaid *pinjar* into ingots, they make it plates, none of which ought to be lighter than 1½ tolas, but a little broader than those which they make in the *Saloni* process. Then having besmeared them with sesameoil, they strew the above mixture over them, giving them for every strewing two gentle fires. This operation they repeat three or four times, and if they want the metal very pure, they repeat the process till it comes up to nine ban. The ashes are also collected, being a kind of *Kharal*.

The Method of Extracting the Silver from Ashes

Whatever ashes and dross have been collected, both before and after the process of aloni, they mix with double the quantity of pure lead, put them into a crucible, and keep them for one watch over the five. When the metal is cold, they refine it as described under the article *Sabbak*. The ashes of it are also kharal. The saloni process is also performed in other ways well known to those conversant with the business.

The Paniwar having melted the *Kharal*, separates the silver from the copper. His fee for every tola of silver is 1½ dams. As a return for the profit he makes, he pays monthly 300 dams to the diwan. Having reduced the *Kharal* to small bits, he adds to every man of it 1½ sers of tangar (borax) and three sers of pounded natron, and kheads them together. He then puts this mass, ser by ser, into the vessel above described, and melts it, when lead mixed with silver collects in the pits. This is afterwards refined by the process of the *sabbak*, and the lead which separates from this, and mixes with the ashes, turns punhar.

The Paikar buys the saloni and kharal from the goldsmiths of the city, and carries them to the mint to be melted and makes a profit on the gold and silver. For every *man* of saloni, he gives 17 dams, and for the same quantity of Kharal 14 dams, to the exchequer.

The Nicho-i-wala brings old copper coins, which are mixed with silver, to be melted; and from 100 tolas of silver 3½ rupees go to the *diwan*; and when he wishes to coin the silver, he pays a fixed quantity for it as duty.

The Discovery of an Alloy in Silver

Silver may be alloyed with lead, tin and copper. In Iran and Turan, they also call the highest degree of fineness of silver dahdahi, in Hindustan, the *sayrafis* use for it the term *bist biswa.* According to the quantity of the alloy, it descends in degree, but it is not made less than five and no one would care for silver baser than ten degrees. Practical men can discover from the colour of the compound, which of the alloys is prevailing, whilst by filing and boring it, the quality of the inside is ascertained. They also try it by beating it when hot, and then throwing it into water, when blackness denotes lead, redness copper, a white greyish colour tin, and whiteness a large proportion of silver.

The Method of Refining Silver

They dig a hole, and having sprinkled into it a small quantity of wild cow dung, they fill it with the ashes of mughilan wood, then they moisten it, and work it up into the shape of a dish, into this they put the adulterated silver, together with a proportionate quantity of lead. First, they put a fourth part of the lead on the top of the silver and having surrounded the whole with coals, blow the fire with a pair of bellows, till the metals are melted. Which operation is generally repeated four times. The proofs of the metal being pure are a lightning-like brightness, and its beginning to harden at the sides. As soon as it is hardened in the middle, they sprinkle it with water, when flames resembling in shape the horns of wild goats, issue from it.

It then forms itself into a disc, and is perfectly refined. If this disc be melted again, half a surkh in every tola will burn away, i.e. 6 mashas and 2 surkhs in 100 tolas. The ashes of the disc, which are mixed with silver and lead, form a kind of litharge, called in Hindi *Kharal* and in Persian kuhna, the use of which will be hereafter explained. Before this refined is given over to the Zarrals, 5 mashas and 5 surkhas are taken away for the Imperial exchequer out of every hundred tolas of it, after which the assay master marks the mass with the usual stamp, that it may not be altered or exchanged.

In former times silver also was assayed by the banwari system; now it is calculated as follows: in Iraq and Khurasan, and the lari and misqali. Which are current in Turan, there are lost three tolas and one surkh, and of the same quantity of the European and Turkish nirjil, and *mahmudi* and *muzaffari* of Gujrat and Malwa, 13 tolas and 6½ mashas are lost, they become them of Imperial standard.

The *Qurs-Kub* having heated the refined silver, hammers it till it has lost all smell of the lead. His fees for the weight of 1000 rupees is 4½ dams.

The *chashnigir* examines the refined gold and silver and fixes its purity as follows: Having made two tolas of the refined gold into eight plates, he applies layers of the mixture as above described, and sets fire to it, keeping out, however, all draught, he then washes the plates, and melts them. If they have not lost anything by this process, the gold is pure. The assay-master then tries it upon the touchstone, to satisfy himself and others. For assaying that quantity, he gets 1-2/5 dams. In the case of silver, he takes one tola with a like quantity of lead, which he puts together into a bone crucible, and keeps it on the fire till the lead is all burnt. Having then sprinkled the silver with water, he hammers it till it has lost all smell of the lead; and having melted it in a new crucible he weighs it, and if it has lost in weight three birinj (rice grains) it is sufficiently pure, otherwise he melts it again, till it comes to that degree. For assaying that quantity, his fee is 3 *dams*, 4½ *etals.*

The Niyariya collects the Khak-e-khalis and washes it, taking two sers at the time, whatever gold there may be amongst it will settle, from its weight to the bottom. The khak, when thus washed, is called in Hindi *Kukrah*, and still contains some gold, for the recovery of which, direction shall hereafter be given. The above mentioned adulterated sediment is rubbed together with quicksilver, at the rate of six *mashas* Quicksilver per ser. The quicksilver from its predilective affinity, drawns the gold to itself, and forms an amalgam which is kept over the fire in a retort, till the gold is separated from the quicksilver.

For extracting the gold from this quantity of Khak, the *Niyariya* receives 20 dams, 2 jetals."

13

Fine Metal-Work

The term 'fine metal-work' describes a type of craftsmanship well recognized but difficult to define. It falls within the province of technology as well as that of art.

The metals available to the ancient metallurgists were gold and silver (frequently as the natural gold-silver alloy electrum copper, tin, lead and iron. When knowledge of the working of the more refractory metals was attained, iron and bronze were used for decorative work. Silver ores might have been obtained from Rajasthan, Afghanistan, Armenia, and Persia. The source opened to the miners for copper ore were Rajasthan and Afghanistan and lead from Ajmer area. The Harappan metalsmiths knew the technique of bead making, soldering, sheet making, rivetting, coiling and cire-perdue casting—lost wax metal casting process. Many of the technical processes in use today are essentially the same as those employed in these days. The early metal worker was familiar, for example, with hammering, embossing, chasing, inlaying, gilding, wiredrawing, and the application of niello, enamel and jems.

Working in Sheet Metal

The early craftsman developed the technique of forging native copper. The question whether the hot-forgiving or cold-working process was used by ancient smiths can be answered only from knowledge of their tools. Hot-forging implies efficient tools for holding the work during hammering. A modern blacksmith's tongs are about 20 inch long, the jews extending some 3 inch beyond the hinge. His grip when holding them equals a pressure of atleast 20 kg. Suppose this pressure to be applied 16 inch from the hinge of the tongs, and the grip on the work to be centred about 2 inch beyond the hinge, then the grip on the work will be 20x16=320 or $20x8^2=320$ Kg with this ample hold the work does not slip, but it will be held about 18 inch from where the hammer-blows fall. No metal-worker would attempt to raise or sink a bowl with so distant a grip.

Consider the fabrication of a vessel, of a single original piece, or of a number of pieces soldered or riveted together after each had been given its particular form. The metal disk from which the vessel was to be constructed had first to be prepared. The procedure for preparation of a flat sheet from the rounded button of bronze. He first flattened it with a heavy hammer and annealed it. Then he beat it with a convex-faced hammer, the face in this case measuring ½ inch by about 1/8 inch blows from such a hammer stretch the metal at right-angles to the longer side of the hammer face. Without rotating it, the smith worked all over one surface of the cake. He then turned the cake over and gave it a series of blows, this time in a direction at right-angles to the first series. Thus he stretched the metal evenly in two directions at right-angles, so that the thinning proceeded regularly. He would anneal it at intervals when it hardened. Finally, if a smooth sheet was required, he would beat it with a flat-faced hammer to remove the dents.

If the smith wished to produce a bowl, he had a choice between two ways of working. He might sink it from a small, comparatively thick disk by continued hammering on the inside of the bowl, or raise it from a thinner, larger desk by hammering the outside of the bowl while it was supported locally by a suitably shaped 'Stake' or anvil. Stakes are made in a great variety of forms, so that a work of any profile may be shaped upon them. To sink a bowl, the smith would hold the disk on an anvil of metal and, with a convex-faced hammer, beat it at or near the centre. Each blow would thin the metal a little as the rounded-face of the hammer compressed the metal and drove some of it literally all around the surface of contact. He would strike a series of blows in a long spiral from the centre of the disk, but would stop hammering before he reached the edge. This hammering would produce a shallow dish which would require annealing. The smith would repeat his hammering again and again always beginning near the centre and stopping before he reached the edge, and annealing the work before proceeding further.

When the smith had produced a bowl of the general dimensions desired, he would employ a suitable stake, conveniently curved, on which the planish it. Planishing is a general, all-over hammering to bring the surface to a smooth, even finish. It is carried out with a flat-faced hammer. A certain amount of final shaping proceeds simultaneously. It should be remembered that whenever a piece of metal is planished, the blows must stretch it a little continued hammering on any supported area of the bowl will cause an expansion of the work at that spot, and if long continued will produce a bump there.

Should the craftsman decide to raise the bowl instead of sinking it he would find it convenient to use a raising-hammer. In its most useful form, this tool has two faces, each measuring about 1¼ inch by rather less than 1/4 inch in thickness. The face in general use in flat, but its front corners and all its edges are rounded away so that they cannot cut the metal. The second face, though as wide as the other, is convex instead of being flat. Its edges to right and left are ground off like those of the other face. It is employed to thin the metal, should the bowl become too thick-walled rather than tall enough. A thinner hammer, as wide as the other and similarly shaped, but only 1/8 inch in thickness, may be employed to raise a bowl from unusually thin metal. In order to understand the technical achievement in the execution of such a work we must consider what is involved in making even a simple bowl.

To raise a bowl, say 6 inch high, 5 inch across the mouth and 3 inch at the base, with either straight or curved sides, one may take a disk of metal 1/20 inch thick, and about 10 inches in diameter (i.e. the height plus the average diameter of the bowl which is to be made). The stake on which the raising is to be done may be a straight bar of iron or brass, or even of hard wood, say 2 inches in diameter, fixed horizontally, or a T-shaped stake with one arm of about that size. The end of the stake is generally under cut by about 20°, so that it shall not impede the free movement of the bowl while it is being raised. One could raise this bowl on a stake or bar only an inch in diameter, but the small radius of such a bar would make the work more difficult. A circle about the size of the base of the bowl is scratched lightly round the centre of the disk. The sheet of metal is held against the end of the stake, with the scratched circle in line with the corner of the stake, and the further edge of the metal is tilted up about an inch above the top of the stake. A sharp blow from the hammer 1/4 inch beyond the scratched line will drive all the metal touched by it into close contact with the top of the stake. The blow must be heavy enough to bend the metal sharply at an angle to its original direction, and the displaced metal must be driven downwards as far as it will go. The disk is now rotated a little, and a second blow is delivered along side the first. The metal is again rotated and struck, and the hammering is repeated right round the circle and continued in a long spiral till it reaches the edge. By now, the metal will have taken a saucer shape and have become hard. It must be annealed before the hammering is resumed, and again at the conclusion of each subsequent course. The work proceeds until the mouth of the bowl is of little less than the required diameter, and the desired height has been reached.

The utmost care must be taken to avoid the formation of a sharp fold, for it usually becomes a crack. Should a pleat or fold seem to be developing, the point of the pleat, after annealing, should be carefully tapped until it is possible to flatten out the incipient defect. If at any time the metal seems to be wearing too thin at the angle where it rests against the corner of the stake, the raising should be continued from another circle at a little distance above the original line. If the shape of a bowl seems to require it, the latter part of the raising may be done from an entirely new circle, perhaps half-way up the profile of the bowl. A certain amount of angularity is inevitable in raising, but it causes no inconvenience, for, when the mouth of the bowl has attained approximately correct dimensions, the angularity may be obliterated in the final shaping and planishing. During the shaping, all irregularities in the profiles of the bowl which may have developed during the raising are hammered out, whole the bowl rests temporarily on any stake which fits that particular part of its curve. The planishing which follows eliminates any remaining blemish.

Embossing or Repousse

Embossing (or repousse) is the art of raising ornament in relief from the reverse side executed with hand-controlled punches. In modern practice, the use of the term repousse has been extended to cover also that work done on the face of the metal to which the term chasing strictly belongs. One of the ways of decorating a piece of metal was to raise a pattern in relief, by blows on its underside while its decorated side rests on a fielding bed of wood, lead, or pitch. Much depended upon the character of relief. If there is to be an all-over pattern having no relief, it is simplest to fix the metal on a bed of pitch by pressure or slight heat, and work over the whole pattern from the front with a tracer, this is a small chisel with a blunt edge. The grip of the pitch keeps the surface flat, and prevents any disturbance of it by the concentrated local pressure from the hammer through the tool. If we consider the groove produced by the traces, we shall find that the metal at the bottom of the groove has been compressed, and that some of the material which once filled its site has been driven up as a slight ridge on either side. Naturally, these small ridges disappear rapidly under polishing or wear, so that they may not be visible on an object that has been in use. If the metal is thin, a ridge may be driven up on the underside, but if it is thick there will be no evidence of the line there. Traces of the intermittent advance of the tool under the hammer-blows may be left as 'stitches' along the line.

In brief Embossing (or repousse) is the art of raising ornament in relief from the reverse side. The design is first drawn on the surface of the metal and the motifs outlined with a tracer, which transfers the essential parts of the drawing to the back of the plate. The plate is then embedded face down into the asphalt block and the portions to be raised are hammered down into the yielding asphalt. Next the plate is removed and re-embedded with the face uppermost. The hammering is continued this time forcing the background of the design into the asphalt. By a series of these processes of hammering and re-embedding, followed finally by chasing, the metal attained its finished appearance. There are three essential types of tools—for tracing, for borring, and for chasing—as well as a specialized tool, a snarling iron or spring bar, which is used to reach otherwise inaccessible areas. Ornaments in relief is also produced by mechanical means. A thin, pliable sheet of metal may be pressed into moulds, between dies, or over stamps. All of these methods have been known from antiquity.

Chasing

Chasing is the term used to describe surface decoration carried out on the front or outer surface of a piece of cast or wrought metal by means of tools, either 'tracers' or 'punches'. These are of simple shape and hand-directed. In general, they are employed to outline or to complete the modelling of, the forms

To produce a pattern with relief of not more than, say 1/16 inch take a smooth evenly grained board, rather larger than the metal and fasten the sheet of metal down all around by nails with flattened heads, placing them about 1/4 inch beyond the edges of the metal. Bend over the heads of the nails to hold down the metal all round without injuring its surface. With a flat-ended punch of hard wood, say 1/2 inch square and 4 inch long, drive down into close contact with the board that part of the metal upon which work is to be begun. Then work the pattern with the tracer. When the metal shows signs of rising off the board from hammering, it may be tapped back again with the wooden punch. The design having been outlined, the background is driven down with chasing tools, melted if desired. Thus the pattern stands up in a very low relief.

If a higher relief is required, another method may be adopted. Consider a relief of which the highest part is about as far above the ground as its own diameter, say a head or figure almost in the round. The surface of a sphere, from which we could work the head, has four times the area of a flat disk of the same diameter, so that the part of the original plate from which the head is to be formed would be stretched to a quarter of its original thickness when punched into the roughly spherical head. If with a round-headed punch we drive up from the back a hemispherical bars on a flat metal sheet, the greatest stretching will take place near the top, and if the hammering is continued then it will eventually crack there.

To avoid this mishap, the metal must be annealed at intervals. It is important, moreover, that the hammering and consequent thinning should be distributed all over the worked area. As the upper part of the bars becomes thinner, its sides will have suffered little stretching, being merely bent from the horizontal to the vertical. These parts must then be thinned to the vertical. These parts must then be thinned to provide material for the sides and back of the head. Therefore, after driving upwards sufficient material for the front of the head, the background and the sides of the bars are driven downwards and inward by work from the front. This work is continued, using any conveniently shaped tools, until the ball-like projection is of approximately even thickness. From it, the head may then be formed. By working in this way, a figure in high relief, say 2 inch long, and almost in the round, may be hammered from a sheet of silver 1/100 inch thick.

In brief chasing was accomplished with hammer and punches on the face of the metal. These punches are so shaped that they are capable of producing any effect—either in intaglio (incising beneath the surface of the metal) or in relief that the metal worker may require. The design is traced on the surface, and the relief may be obtained by beating down the adjacent areas to form the background. Such chased relief work sometimes embossed work, but in the latter process the design is bossed up from the back.

The detalled finish of embossed work is accomplished by chasing, the term is applied also to the touching up and finishing of cast work with hand-held punches.

Stamping

When many identical pattern in relief were required for a necklace or a bracelet, a craftman might turn his thoughts to inventing a more speedy method than making each panel individually by repousse. He would soon find that by chasing, filing and grinding the design on the face of a bronze or iron punch he had acquired a tool with which he could produce as many copies of the design as were needed, without further labour. He would employ the punch in conjunction with a cake of lead. With a few strong hammer-blows he would make a deep impression of the punch-face in the lead. Then he had but to lay over the depression of thin sheet of gold or silver, properly annealed, and drive it in with a few steady blows. The relief thus produced would be an exact copy of the pattern on his punch. To finish the work he had but to trim the metal round, and file it true.

Occasionally the craftsman set to work in another way. With steel tools he would carve the design in intaglio in a piece of bronze. During carving he would take an impression in wax from time to time, to see how the work progressed. When the die was complete, he would take a piece of thin sheet metal, properly annealed, and lay it on the die with a piece of thick sheet lead above it. A few blows with the hammer would give a complete reproduction of the design, which he could trim and finish at leisure.

Engraving

Confusion has resulted from indiscriminate use of the term engraving. From Palaeolithic times flint points were used for cutting patterns on bone, ivory and soft stone. Such works may properly be described as engraved. Later marks were occasionally scratched on metal with flints, but at no time were regularly cut patterns on metal effected with flint tools. There was no art of engraving on metal in the Bronze Age, or indeed until steel tools became available.

Experimental attempts with bronze tools, variously hardened to engrave upon gold, copper, and bronze objects prove that no regular engraved work can thus be done.

The Urarartu (people of Ararat) and the Assyrian seem to have been the first to arm their warriors with iron. To this equipment was partly due the great extension of Assyrian power, even to the highlands of Armenia in the north and the borders of Nubia in the South, under Ashur-nasir-pal II (884-859 B.C.) and his successors.

Engraving is to cut or incise a line. Engraving is always done with a cutting tool, generally by pressure from the hand. If detaches material in cutting when pressure is applied with a hammer, the process is called carving. Many art objects of medieval period show that engraving was practiced with perfection.

Soldering and Welding

An insistent problem for the early metal works was the

joining of separate pieces of gold or copper. They could be fastened with pins or rivets, which were indeed commonly used for fixing a handle to a dagger or knife, or for sheet metal work, as in a type of vessel the body of which was built up of separate pieces. It is possible to join pieces of metal at the comparatively low temperature of 120°C, or even less.

There has been confusion in technical literature as to the methods actually employed by the early craftsmen in joining pieces of gold, electrum, silver, bronze, or copper. Statements that they were either fused together by autogenous welding, without solder, or welded together by the hammer as a blacksmith welds iron, are erroneous. The soldering of gold and silver was known to Indians even during antiquity. The ancient goldsmith, to whom so easy a process as soldering was available, would not have attempted the difficult, if not impossible task of welding.

The end of a piece of wire heated to fusion will suddenly run into a sphere, owing to the surface tension. If one lays two lengths of wire end to end over overlapping a little, the two do not run together, but each draws back when melted. So, too, if adjacent edges of two plates of metal are fused they will draw back from one another. When, however, a more fusible material is present, it forms a bridge between the two, alloying itself with each, running through the joint, and fastening the parts together.

When copper or any of its alloys is heated, as in the charcoal furnace—the only one available in antiquity—a film of Copper oxides forms on the surface. This keeps a heated sheet of copper from direct metallic contact with another. Neither a flux nor hammering will unite the surfaces. Copper surfaces, unlike those of iron, could not be welded together by heat or hammering. Early metal workers knew soldering and employed it widely.

Gold behaves in a rather exceptional manner. So in order to make a little box, he beat out some almost pure gold into strips from which to form the sides. He allowed their ends to overlap. Perhaps a quarter of an inch, and burnished each extremity down to that parts of the strip against which is tested, with the result that it adhered it firmly. In the interval between the two burnished (welded) joints a double thickness of gold may be seen. A part from such adhesion by burnishing, no other welded joint in gold is known from the ancient world.

Fluxes: Fluxes were used with nearly all solders. A surface of heated metal exposed to the air is liable to become oxidized, and the film thus produced will interfere with the flow and adhesion of the solder. It is therefore usual to employ a flux which, running through the joint, excludes the air and keeps the surface clean. Borax is employed as a flux for many hard solders, and zinc chloride for soft.

Sweating: When the contiguous surfaces of two or more pieces of metal have been separately coated with some more fusible material, such as a solder, and the temperature is raised high enough to melt this coating, a joint is formed which may be described as sweated e.g., a joint may be sweated between two pieces of a tin canister (made from tinned iron) without the aid of any additional solder.

Burning together was practiced from the Bronze onwards. By it, a joint can be made on a bronze tool or weapon without the aid of solder. A bronze sword, broken at the hilt, might be thus repaired. The smith fitted the pieces together and formed a mould in clay round them. He left a passage all round the joint, and provided the mould with a funnel—shaped pour for the introduction of the metal, and an overflow hole. Then he poured into through the mould large quantity of molten bronze. The metal flowed between, and heated up, the broken parts of the sword, partly melting them. Most of the molten metal escaped through the overflow hole, but enough remained to make the joint strong. Superfluous metal could be cut away later. Similarly, a new part could be cast on to an existing metal object. Care was always needed that sufficient molten metal passed through the mould to fuse the old work, the oxides rise to the surface as a scum.

Niello

Niello is the process of inlaying engraved ornamental designs with niello, a silver sulfide or mixture of sulfides. The first authors to write on the preparation of niello and its application to silver were Eraclius and theophilus in or about the 12th century, and Benvenuto Cellini, during the 16th. According to each of these authors, niello is made by fusing together silver, copper, and lead and then mixing the molten alloy with sulfur. The black product (a mixture of the sulfides of silver, copper, and lead) is powdered; and after the engraved metal, usually silver, has been moistened with a flux (a substance used to promote fusion), some of the powder is spread on it and the metal strongly heated; the neillo melts and runs into the engraved channels. The excess niello is removed by scraping until the filled channels are visible, and finally the surface is polished.

14

Bidri Metallurgy

Bidriware is a metallurgical compound, and the articles made from it are greatly admired for the elegance of their form and the gracefulness of the pattern with which their surface is covered. The Zinc-Copper alloy of Bidriware form an excellent dark background for the shining silver or gold inlay. It never rusts or corrodes, but is brittle and liable to break if dropped. The craft of Bidriware is a kind of damascene work. The name damascene seems to have been derived from Damascus, the original place where it was carried to perfection in the ancient days. It is a system of ornamentation consisting of encrusting one metal on the other. Bidriware is admired for the exquisite decorations in the articles of every day use, like Huqqa bottoms, jewellery box, spitoons, goblets, jugs, wash basins etc.

Damascening and encrustation both denote the surface ornamentation of one metal by the application of others. In damascening (Koftgari) iron or steel in usually ornamented with gold or silver wire. In the various forms of encrusted work, the ground metal is rarely used and the applied metals are seldom in the form of wire. The art of damascening appears to have originated with the ornamentation of swords and other weapons.

The art with silver or gold is called Bidriware—so called after the name of the old city of Bidar, which with its historic tombs and a vast imposing fort, is a monument of history. The origin of Bidriware is not very well known. However, the name is associated with Bidar in Deccan and its history dates back nearly 400 years to the Bahmani and Baridi Dynasties under whose patronage it flourished. It reached its greatest perfection towards the end of the rule of these dynasties. Bidar was a part of the erestwhile Hyderabad State till the date of States Reorganization in 1956, and is at a distance of 83 km. from Hyderabad city. Now Bidar is in Mysore State. It is situated on 17°-55′ N and 77°-32′E on an elevated and healthy plateau 2,330 ft above the mean sea level. The place provided an ideal seat for the famous craft known throughout India and also abroad as Bidriware.

According to local tradition, the Kakatiya Rajas of Warangal endowed a temple of Mahadeo, which was here, a town sprang up in its vicinity in the mid-thirteenth century. Bahman Shah Gangu, the first Bahaman King of Gulbarga took Bidar. In 1430 A.D. Ahmed Shah Wali the tenth Bahmani king founded the modern town, built the Fort and removed the capital here from Gulbarga on the dissolution of the Bahamani kingdom. It fell to the Bariad Shahis of Bidar' who reigned from 1492 to 1609 A.D. afterwards the Adil Shahis of Bijapur annexed it. The city was plundered by Malik Ambar, the Nizam Shahi minister of Ahmednagar in 1624 A.D. but was recovered by the King of Bijapur and remained part of his realm till 1656 A.D when Aurangzeb annexed it.

The town remained in the possession of the Mughals till the first of the Nizams declared their independence early in the 18th century. The town of Bidar must have been prosperous as it appears from its palaces, mosques and other buildings.

Bidar is celebrated amongst other things, for the manufacture of a species of metalware which is styled Bidriware. A large number of articles of Bidriware were made expressely for presentation to H.R.H the Prince of Wales when he visited India in 1875, and they now hold a post of honour among the collection of Indian curiosities by this Royal Highness.

It is difficult to give an authentic and continuous history of Bidri Art. According to the local craftsmen, sometimes about 900 years ago the craft was brought down to India by migrants from Iraq. The original home of the work seems to be Iran, but it travelled to Ajmer in Rajasthan through migrants. After many years on Abdullah-bin-Khaiser, who was conversant with the knowledge of this art came down to Bijapur and settled there. He applied the art only for his sustenance, working on it now and then and selling the articles. It is said that a carpenter by name Sivanna, who won recognition for his services and gained confidence of Khaiser, was taught the craft. Tuhs Sivanna became the first native of India to learn this art. Thi must have been about 500 years ago. That remained a sec et in their family and Sivanna's sons and descendants produced exquisite articles for presentation to Nawabs and nobles.

The beautiful articles presented to Allauddin II (1434-1457 AD) of Bidar on the occasion of his coronation impressed him very much and he invited the craftsmen of Bijapur to settle at Bidar itself. The workers were provided with all facilities and palace authorities looked after their comforts. The workers used to prepare beautiful huqqa base, slabchis, aftabai and cot legs for the palace, some of which are said to still preserved in the local museum at Bidar with the name of the craftsman Yellana engraved on them. The craft was named Bidriwork by the King at the time of his birthday celebrations. His son Ahmed Shah Bahmani had an intense liking for the art and it flourished well under his bountiful patronage. It is not known by which name the craft was known before it was named Bidriware by the Sultan of Bidar. It is said that the craft received encouragement and patronage from the Chief Minister Mahmud Gawan of Bidar, who was himself a great connoisseurs of art and patron of literary men and artisans. The Russian traveller Althanasins Nikitin, who travelled in the Kingdom between 1470-1474 AD took with him the beautiful Bidri articles for presentation to the Russian Emperor.

The Bidriware has 'common ancestry with the art practised in Arabia and Persia of inlaying gold and silver on steel and copper. From that it differs in the base material, which is in this case is an alloy of Zinc and Copper and sometimes an alloy of Zinc with small proportion of other non-ferrous metals of a dull leaden colour.

Like other Persian utensils of metal with scripts and design produced by inlaying gold and silver. This particular type of work probably developed by experimenting with various alloys to guarantee brilliance by contrast.

Raw Materials and Tools

The Chief raw materials were, of course, the metals in the proportion of Bidriware articles, of which Zinc was used in bulk quantities. Zinc and Copper mixed in the proportion 16:1 for the base material for all the article. The important raw materials required for the manufacture of Bidriware were: Zinc, Copper, Red Clay, Resin, Castor Oil, Sangjeera (white stone powder), charcoal, Bees wax, silver, gold, lead, tin, groundnut oil.

Tools and Implements

The tools and implements used in the manufacture of Bidriware articles were mostly simple. The tools used were very similar to those applied by the goldsmiths and silversmiths, as the inlaying work involved was similar to those crafts. In mould making and casting the required article was first done in a process similar to that adopted in making brass castware articles and for this purpose there were hollow metal frames (darza boxes of flasks) of square or rectangular shaped with four walls of two inches width on the four sides, a furnace specially constructed with an attached blower, crucibles etc.

Process of Production

The desired vessel was first moulded and then reduced to the exact shape on the turning lathe. It was next engraved or chased in varying depth according to the quality or kind of work that was required to be produced. It was, therefore, smoothed, polished and coloured to a dark green or black colour by ammonia and saltpetre moistened in castor oil and thickened with charcoal. The vessel was slightly heated by placing it in the sun, then rubbed with the colouring preparation and allowed to cool for several hours. It was therefore washed and rubbed with a little oil, when the colour became permanent. There were various forms of Bidri according to the depth of embedding silver and gold and the quality of the metal affixed to the surface. The varieties of workmanship of the design consisted tara kashi (inlay of wire) taihnishan (inlay of sheet), zaranishan (low relief), zarabuland (high relief) and aftabi (design in overlaid metal sheet).

The entire technique of casting can be divided into two broad phases as follows:

1. *Mould Making and Casting*: A mould of the required

articles was made with clay and molten metal was then poured into it. The surface of the cast article was smoothened on moving lathe and with files etc.

2. *Engraving and Inlaying*: The design was drawn with free hand drawing on the surface, engraved with sharp chisels in varying depths and then silver wire or sheet were embedded on the chased pattern by hammering and then the surface was levelled.

3. *Finishing*: The article was blackened and then polishsd carefully.

The two outstanding aspects of Bidriware are:

1. The fanciful designs and patterns that are skilfully wrought with radiant silver or gold and
2. The black background surface of the article which does never rust or corrode.

These two features have earned for this metalware articles undying fame sustained through centuries.

Preparation of Alloy

The components of the Bidriware alloy were in the ratio of copper sixteen parts lead four parts and tin two parts which were melted together and to every three parts of the alloy, sixteen parts of spelter i.e. of Zinc were added. Then this mixture was heated for use. In order to produce the black colour which brought out the pattern, the composition was dipped into a solution of Sal-ammoniac, salt petre, common salt and blue vitrol, a mixture of resin and bees wax being introduced to prevent calcination. The liquid was then poured into a mould made of backed clay and the article thus roughly shaped was turned in a lathe. Artists then inlayed upon it floral or other ornaments of silver or gold in fanciful patterns.

In order to bring out the important steps in the process of production, the procedure followed in respect of two articles, namely a decorated plate and a goblet of round shape is described here. Mould making and casting a decorated plate:

The pattern of the plate, which was to be cast and which has been prepared earlier, was brushed clean with a hair brush especially the engraved design on it.

On the zinc sheet (12″ sq.) the above pattern plate was placed in the centre. Sangjeera powder which was earlier tied up in a piece of cloth, was dusted on the Zinc sheet including the pattern plate, to the extent that it formed a thin white layer all round. This prevents the clay from sticking on to the pattern and to the Zinc sheet, which were removed at a later stage before casting the plate. There were hollow frames made of iron sheets which were called darza boxes. These were of various sizes. Half of the darza boxes which was bigger than the pattern plate was then kept on the zinc sheet in such a manner that the pattern plate was in its centre. This was the upper half darza box. The fine screened red clay with resin and castor oil mixed in it was then laid in a thin on the pattern plate and near about. The reason for using this fine clay was to get smooth surface in the cast articles and secondly to avoid cracks which sometimes formed if only dry clay was used. Over this layer, layers of ordinary red clay was laid one after the other till the whole darza box was completely filled in.

The artisans then thrust this clay with pressure so that it filled in the frame tightly. Later, the extra clay was removed from the darza with the help of an iron blade. The clay was levelled with the level of the upper edge of the darza box frame. Over this was placed a wooden plank big enough to cover the darza box as a whole.

The whole frame was then lifted and reversed i.e. the zinc plate now formed the upper side with the wooden plank on the floor. The former was removed from the darza box and Sangjeera powder was dusted all over. The only precaution that was taken was that the engraved design was dusted lightly that it did not fill in all the depth and that no engraved portion was left undusted. That was the reason why very fine Sangjeera powder was required for dusting. All the surface looked white with a uniform layer of this powder.

The other half of the darza box frame was then kept on it with the frame locks in perfect contact. An iron pipe which was equal to the height of the frame in length and about 0.75” in diameter was placed in a position. Just in line with the pairing spout of the frame and in the opposite and leaving on inch's space from the frame when casting took place. That pipe formed the ingress channel for the molten alloy. The pipe was also dusted all over with the sangjeera powder, as it was removed from the darza box at next stage and it was allowed not to stick on to the clay, which was filled in along its sides and pressed too. Now over the pattern plate and near about it, a layer of fine resin and castor oil mixed with red clay was laid followed by layers of ordinary red clay thrust with pressure tightly as was done in preparing the other half of the darza box. The two halves of the darza box were then unlocked and kept separately. One could clearly notice that the engraved design had attained the embossed effect on the upper half of the darza box. In this process the iron pipe had been removed leaving a round hole in the upper half of the darza box. The hole so formed was dusted with sangjeera powder and was smoothened with thumb.

The projection of the hole was marked on the other half of the darza box and keeping it on the vertex point two straight line channels were cut with an iron blade up to the outer ring of the pattern plate, thus forming a triangle with the curved edge of the plate. The channel was about 0.25" deep and was circular in form with utmost dexterity and with the help of finger nails the pattern plate was lifted right upwards, which responds fully. When the two halves of the darza box were rejoined together the hallow space vacated by the plate formed the mould for the article to be cast with molten metal. Sangjeera powder was sprayed on the channels cut earlier and with the help of right hand thumb's outer edge. They were further smoothened and rounded.

As the edge of the thumb was convex in shape. It gave a concave effect on the soft clay. The thumb moved from the plate mould's edge towards the vertex point. This channel acted as a stopper, and when the molten alloy was poured through the round hole of the upper half of the darza box, it drained right to the hollow mould formed by the pattern plate through the V-shaped channel.

The halves of the darza box were once again placed in their original position one above the other, the upper wooden plank was replaced with a small sized plank just enough to cover the breadth of the box and the mould of the pattern plate completely.

Simultaneously along with the preparation of the clay mould the metal alloy was melted in the furnace nearby in a crucible. The alloy was composed of 16 parts of zinc and one part of copper or proportion of 94:6. First the zinc was well melted in the crucible and then the copper was added to it. When the alloy was ready for casting one could notice the blue flame coming out of the crucible. In the right hand the hot crucibles with the molten alloy was firmly held with tongs and the alloy was passed into the round hole, while with the left hand the wooden plank was pressed, so that the alloy did not spread out of the two walls of the moulded plate.

The darza box was again unlocked and cast plate was lifted out with the help of long handled pliers held by the left hand, and with the plate to dust out the clay with which it has adhered to it.

The cast article was then smoothened to the desired shape and fineness with the lathe machine and also with files. Normally, the concave side of the plate was smooth but the convex side required heavy filling. The plate was placed against a wooden block in reverse position i.e. with the convex surface upward, whilst the lower end of the plate was held in place by means of clamping the toes of left foot on it. A flat file was held in the right hand and with the pressure of the left hand laid on its end, the filling was done by first thrusting the file forward and upwards and then backwards. In this way the whole convex surface was smoothened.

After passing through these stages the cast article assumed the shape and size of the pattern piece. Thus, the first stage of production was complete.

The process described in these paragraphs was essentially the same for all other articles of flat type.

Mould Making and Casting a Goblet

The red clay was cleaned, sieved and the fine powder was mixed with resin and castor oil so on to make it a thick paste. The paste was made into a ball around a small stick of about ½" diameter. It was made round and round by manipulating with hand so as to fit exactly into the metallic model, the halves of the model being pressed round the clay ball. The ball was smoothened nicely. At the two ends of top and bottom was fixed some clay to cover the ends of the stick which passed through the spherical ball. The component parts of the metallic model of the articles which was to be reduced were used for making the mould. In case of a goblet, the round belly, long neck and the base were prepared separately and afterwards soldered with one another. Usually each piece was divided into two halves vertically or even horizontally, if it was of spherical shape.

A frame of darza box was put on the ground and was filled with clay tightly. In the middle of the frame the spherical clay ball was surrounded by the two metallic halves and pressed with stress so that half of its volume was covered by the clay powder inside the frames. Then the sangjeera powder which was tied up as a packet in a cloth and sprayed in a thin layer over the visible portion of the clay ball and the metal mould. This prevented the clay powder from sticking on to the metal mould and clay ball. Then two more frames were placed over it so as to completely cover up models and they were again filled with the clay powder tightly pressing with hands. After all the space was tightly filled in, the whole set of frames was reversed topsy and turvy. Then the original lower half was taken out and sprinkled with sangjeera powder. Then the frames were re-arranged into the shape of the box with the metal model fully contained inside and pressed heavily. Then a short iron rod of about 4″ length was pierced into the clay so as to push slightly the metal surface inside so that a passage was made for the molten metal to be poured through and then the rod was removed gently. The set of iron frames was divided into two, so as to correspond to the two halves of the metallic model and those two pieces were removed off slowly taking care to see that the surface cavity made by them was not disturbed at all. Then the frames with the clay were re-arranged into the box. There would have occurred a thin round spherical shape surrounding the inner clay ball, comprising the space vacated by the two portions of the metallic model. This space was the exact replica of the metal model, the clay ball being the necessary contrivance to create hollow space inside.

Side by side with the preparation of that mould metal was melted in a crucible on a furnace with a bellow attached to it. When the alloy was sufficiently boiled and was in a liquid state free to flow, the crucible was lifted from inside the furnace, held out with long tongs and poured carefully into hole from the top of the darza box. Pouring was done slowly and steadily so that it did not split. While pouring the metal one stood upon a raised platform because sufficient stress was necessary to keep the arrangement inside intact. The liquid metal occupied all the vacant space around the inner clay ball and solidified within a few minutes. Then the arrangement of darza box was disintegrated after waiting for five minutes and the cast article was taken out with tongs. The inner clay ball was then removed gradually breaking it piercing with thin iron rods. Thus one portion of the spherical belly of the goblet was cast. It was cleared of all the clay and sent to the hand lathe machine for turning to achieve smoothness of the uneven surface.

Coming again to the above discussed plate, we see that a faint impression of the indented design of the pattern was left on the plate so cast and finished, but this shallow chasing or engraving of the design on the plate was not deep enough to inlay wire or sheet to the required degree of fastness. As such the next step was to engrave the design to the required degree of fastness. For engraving purposes the plate had to be held in a fixed position which was only possible by embedding it on the wax surface. For preparing such surface, wax was melted thick on charcoal even and some ash or saw dust was added to it, when that mixture was warm enough it was transferred and laid as a layer of about 1" thickness on a piece of wooden plank large enough for the plate to rest fully on it and immediately the plate was securely embedded in it with the design side upwards. This made it easy for the craftsman to work, as the position of the plate could be rotated and moved with the wooden plank underneath as and when required. The wooden plank was placed on a small teapoy.

The method of bringing down the outline of the design at the time of casting itself was applied only to the decorative plates which has intricate and broad patterns of figures on them, for in respect of all other articles like jewellery boxes, surahis etc.

It was necessary to draw out the designs at first on the surface, before proceeding with chasing or engraving. A sharp end of an iron spoke called kalam (pen) were used for the purpose. The workers were rich in imagination and he drew the required designs on the metal surface. Usually an experienced worker needed no replica of the proposed design from which he had to copy. He could draw off hand straight with the kalam copper sulphate solution was first smeared to the article and this turned the greyish white surface of the alloy-cast article to black. Thus the artisan was able to look at the figures clearly while drawing. The design or pattern thus drawn out from this basis along which engraving was done in the next operation. With finer pointed Chisels engraving in the required depth was now carved on it. While engraving the chisel operated from right to left. The chisel was held in the left hand and the hammer in the right hand. The whole design was chased or curved all along the lines drawn earlier.

Now silver wires were prepared according to the depth of the engraving. For thinning the wire, it was drawn through steel plates with holes of different sizes. An end of the wire was made pointed by hammering it and held with pliers with one hand. It was drawn through the hole of the intended dimension with the other hand. Afterwards the drawn wire was heated on charcoal hearth and beaten slightly so as to retain its size. For Taihnishan and Aftabi workmanship, the silver sheet was quite simply cut in the size of the design that had been traced on it with a prepared design on paper.

The next operation was enbedding the silver wire. The wires were fitted into the design along with engraving done already. The wire was held fast on the engraved design by a flat pointed chisel grasped by the craftsman with left hand and by the right hand it was hammered into the interstices. Later on, direct hammering was done to have the wire embcdded on the surface securely. The same process was used for Taihnishan and Aftabi techniques. For zar nishan and zar buland work, the outline of the pattern was made on the silver sheet, which was very thin. The leaf was then cut into the desired pieces each a little larger than the space was intended to cover. The margin or rim of each was bent over and the cavity thus formed was filled with a piece of soft lead. Afterwards, this was inverted over the space and the margin of the structure pressed into the engraved outline. Lastly, it was hammered or punched all raised so as to cause the surface metal to embrace and fix the applied piece. The process was then finished by the silver leaf being punched or chased on the surface on completion of the desired pattern, similar to that done in the Aftabi technique. This marked the completion of the second stage.

The third stage which was mostly finishing started by filing the above piece with a flat iron file so that the embedded silver and the base materials surface formed a smooth single level. This was done only in the case of Tarkasi and Taihnishan techniques. When the form of the article to be smoothened was not flat but was round, the help of a scraper, which was just a steel plate having one inch bent at the end at right angles and the end sharpened like a knife was used to smoothening the uneven surface of the silver embedded article. The artisan judged whether the article was fully finished or not by rushing the finger all over the surface.

The final stage and the most important characteristic of Bidriware articles was worked out. In order to blacken the base material, salt petre containing clay and sal. ammonia were mixed in sufficient water in the ratio of 30:5 and the solution was boiled in an open iron vessel on a charcoal hearth. The finished article was now given a dip in this hot solution. The zinc alloy turned permanently jet black while the silver retained its brilliance. The affect became very charming with white shining pieces on a black base. This was the climax of the Bidri process.

Finally, polishing of the articles was done. It was done by preparing a paste of fine charcoal powder and groundnut oil. It was applied in a small quantity on the finished article and rubbed with a cloth. This gave a glazed and shining effect to the surface so polished.

Articles normally make during this period were Huqqa base of different sizes and shapes, balls, bells, coconuts and mangoes. They include spice boxes for offering betal and betal nuts of various forms. Water flagons, trays, water bowls with or without lid, cups, tumblers, spittoons, washbasins, ewers, mir-e-fersh (weights) cot legs, cosmetic boxes and bottles of various shapes and forms.

Bibliography

Abul Fazl, *Ain-i-Akbari* 1867-77, Translated into English by H. Blochmann, Vol. II, Asiatic Society, Calcutta.

Agrawal, D.P. 1976, *Ancient Copper Working*: Some new C Dates. J.I.H.S., Vol. II, No. 2.

1971, *The Copper Bronze Age in India*, New Delhi.

Allchin, B. and R., 1968, *The Birth of Indian Civilization* (Pelican).

Allchin, F.R., 1962, *Upon the Antiquity and Methods of Gold Mining in Ancient India*, J. Econ. Social Hist. Orient, Vol. V.

Alwin, Verrier, Agarias, London.

Bannerjee, M.N., 1929, *Iron and Steel in Regvedic Age*, Indian Historical Qrly., Vol. 5.

Bannerjee, N.R., 1965, *The Iron Age in India*, Delhi.

Bhardwaj, H.C. *Aspects of Ancient Indian Technology.*

Bose, D.M., et al., 1971, *A Concise History of Science in India*, New Delhi.

Brown, J. Coggin and Dey, A.K., 1955, *India's Mineral Wealth*, Bombay.

Buchanan F., 1807, *A Journey From Madras* (London).

Caley, E.R., 1964, *Analysis of Ancient Metals*, Pergamon Press.

Campbell, Cap, J., 1842, *Manufacture of Bariron in Southern India.*

Chakraborti, D., 1976, *The beginning of Iron in India*, Antiquity, L.

Charles, J.A., 1967, *Early Arsenical Bronzes—A Metallurgical View.* American Journal of Archaeology, Vol. 71, No. 1, pp. 21-27.

Coghlan, H.H., 1962, *Notes on the Prehistoric Metallurgy of Copper and Bronze in Old World*, Oxford.

Crackoft, W., 1833, *Iron Smelting in the Khasi Hills*, J.A.S.B., Vol. II,

Desch, C.H, 1937-38, *From Meteoric Iron to 1000 B.C. Transactions of the Newcomen Society for the study of the History of Engineering and Technology*, London, Vol. XVIII.

Dunn, J.A., 1965, *Bull, Geo. Surv. of India*, No. 23, Copper, Delhi.

Forbes, R.J., 1950, *Metallurgy in Antiquity*, Leiden.

Friend, J.N., 1926, *Iron in Antiquity*, London.

Garland, H. and Bannister, C.O. 1927, *Anceint Egyptial Metallurgy*, London.

Ghosh, M.K., 1963, *The Delhi Iron Pillar and its Iron*, N.M.L. Technical Journal, Vol. V, No. 1, pp. 31-45.

Gordon, D.H., 1952, *Early use of Metals in India and Pakistan*, J. Royal Anthropological Institute, London.

Gowland, W., 1899, *Early Metallurgy of Copper*, Tin and Iron in Europe, Archaeologia, London.

Greaves, R.H. and Wrighton, H., 1950, *Practical Microscopical Metallography*, London.

Hadfield, Sir, 1912, *Sinhalese Iron and Steel of Ancient Origin* Journal of the Iron and Steel Institute of London, No. 1, pp. 153-158.

Hegde, K.T.M., *Science Today*.

Hodges, H., 1965, Arteffacts, *Early Materials and Technology*, London.

Joshi, S.D., 1970, *History of Metal Founding on the Indian Sub-Continent Since Ancient Times*, Ranchi.

Kangle, R.P. 1963, *Kautilya Arthashastra—A study*, Bombay.

Knox, R., 1963, *Detection of Iron Carbide*, Structure in Oxide Remains, Archaeometry, Vol. 6, pp. 43-45.

Kosambi, D.D., 1963, *The Beginning of Iron Age in India*, J. Econ. and Social Hist. of Orient, Vol. 6, pp. 309-318.

1 65, The Culture and Civilization of Ancient India in Historic Outline, London.

Krishna, M.S., 1951, *Mineral resources of Madras* (Calcutta).

Lal, B.B., 1956, *An Examination of Some Metal Images from Nolanda*, Ancient India, No. 12, pp. 53-57.

Lamberg-Karlovasky, C.C., 1967, *Archaeology and Metallurgical Technology in Prehistoric Afghanistan, India and Pakistan*, American Anthropologist, Vol. 69.

La Touche, T.H.D., 1938, *A Bibliography of Indian Geology and Geography*,

Luard, Capt. C.E., 1907 *Rewah State Gazetteer* Vol. IV, C. (Lucknow).

Marshall, J., 1931, *Mohenjodaro and Indus Civilization*, Vol. I and II, London.

Mooreraft, William and George Trabeck (1819-25) *Travels in the Himalayan Province of Hindustan and the Punjab in Ladakh and Kashmir in Peshawar*, Kabul.

Munn, L., 1936, *Observations and Notes on the Methods of Ancient Gold Mining in Southern India*, Tr. Min. Geol. Inst. India, Vol. 30.

Needham, J., 1954-63, *Science and Civilization in China*, Vols. I-VII, Cambridge.

Neogi, P., 1914, *Iron in Ancient India, Indian Association for the Cultivation of Sciences*, Calcutta. 1918, *Copper in Ancient India*, Indian Association for the Cultivation of Sciences, Calcutta.

Oldham, M.G.S.I., Vol. XIX.

Partington, J.R., 1935, *Origin and Development of Applied Chemistry*, London.

Piggot, S., Prehistoric India to 1000 B.C. Middlesex.

Pittoni, R., 1948, *Recent Researches on Ancient Copper Mining in Austria*, Man, (London), No. 133.

Pleiner, R., 1971, *The problem of Beginning of Iron Age in India Acta Prehistorical et. Archaeologica*, II.

Rahman, A., and B.V. Subbarayappa, A note on the *Native method of bar Iron Production in South India* (Salem Region) IJHS, Vol. I, No. 2, 1966.

Ray, P., 1956, *History of Chemistry in Ancient and Medieval India*, Calcutta.

Rickard, T.A., 1932, *Man and Metals*, New York and London.

Rickard, T.A., 1939, *Primitive Smelting of Iron*, American Journal of Archaeology, Vol. XLIV.

Rosenfield, A., 1965, *The Inorganic Raw Materials in Antiquity*, New York.

Shamashastry, R., 1929 (Eng. TR.), *Kautilya's Arthashastra*, Mysore.

Sankalia, H.D., 1970, *Some Aspects of prehistoric Technology in India*, New Delhi.

Scott, Helenius and Sir Joseph Banes (1790-1801) *Aspects of Technology in Western India vide Indian Science and Technology* in the 18th century. (By Dharampal Ed.) Impex India (1971).

Singer, C., et. al. 1967, *History of Technology*, Vols. I and II.

Smith, V.A., 1905, Indian Antiquary, Vol. 34.

Subrahmanyam, B.R., 1964, *Appearance and Spread of Iron in India.* An Appraisal of Archaeological Data, Journal of Oriental Institute, Vol. 13, No. 4, pp. 349-359.

Sullivan, John, *Tracts upper India*, written in the year 1779, 1720, and 1788 (London 1794).

Shastree, B.G. *Note on the iron ore in the vicinity of Malwa.* Journal of the Bombay Branch of the Royal Asiatic Society, Vol. I.

Thurstan, E., (1975), *Castes and Tribes of Southern India*, Vol. II.

Tylecote, R.F., *Metallurgy in Archaeology*, London.

Vats, M.S., 1940, *Excavations of Harappa*, Delhi.

Verchare, A.M., J.A.S.B., XXXVI, pt. II.

Voysey, H., 1832, *Description of the native manufacture of steel in Southern India*, J.A.S.B., I.

Watts, W.G., *British Garhwal, A Gazetteer*, Vol. XXX of Distt. Gazetteers of the United Provinces of Agra and Oudh.

Water, E.R., 1907, *A Monograph on Iron and Steel Works* in the Province of Bengal.

Wheeler, R.E.M., 1959, *Early India and Pakistan*, London.

Yule, H. and A.C. Burnell, 1903, *Hobson-Jobson* (London).

Index

PLATE 1

Iron Pillar Near the Qutub Minar, 5th Century.

PLATE 2

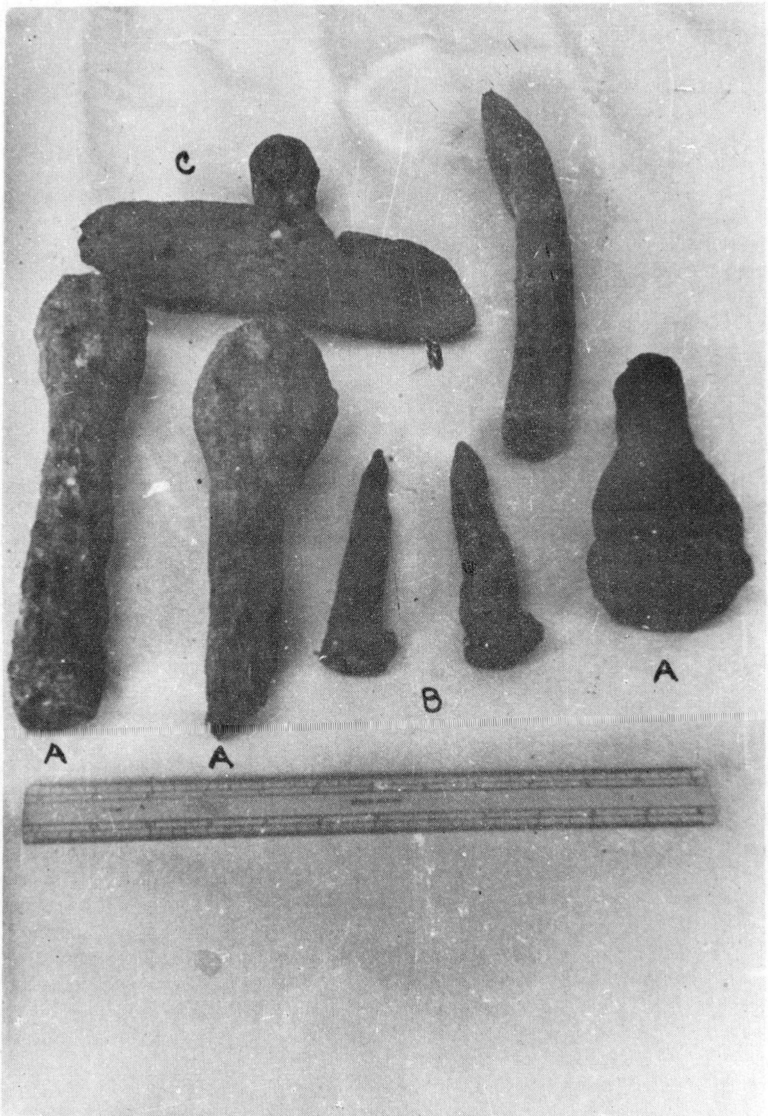

Mining Tools Found Inside Old Workings.
A. Iron Hammer, B. Chisels.C. Spade.

PLATE 3

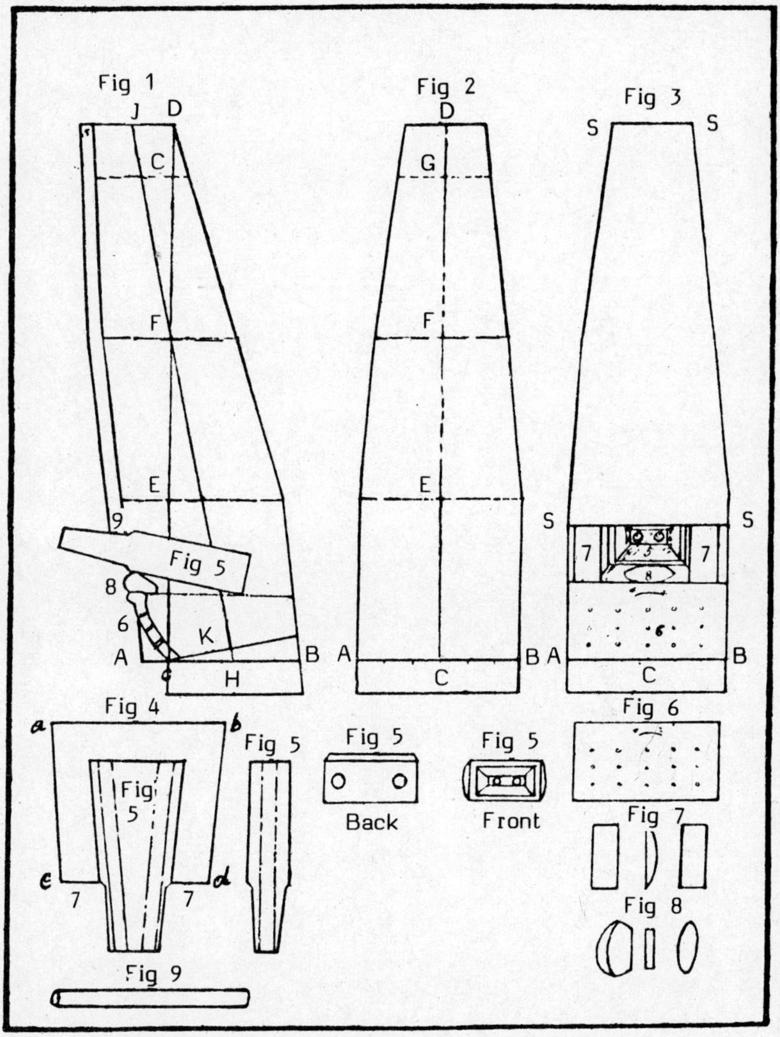

Geometrical Outline of the Furnace.

PLATE 4

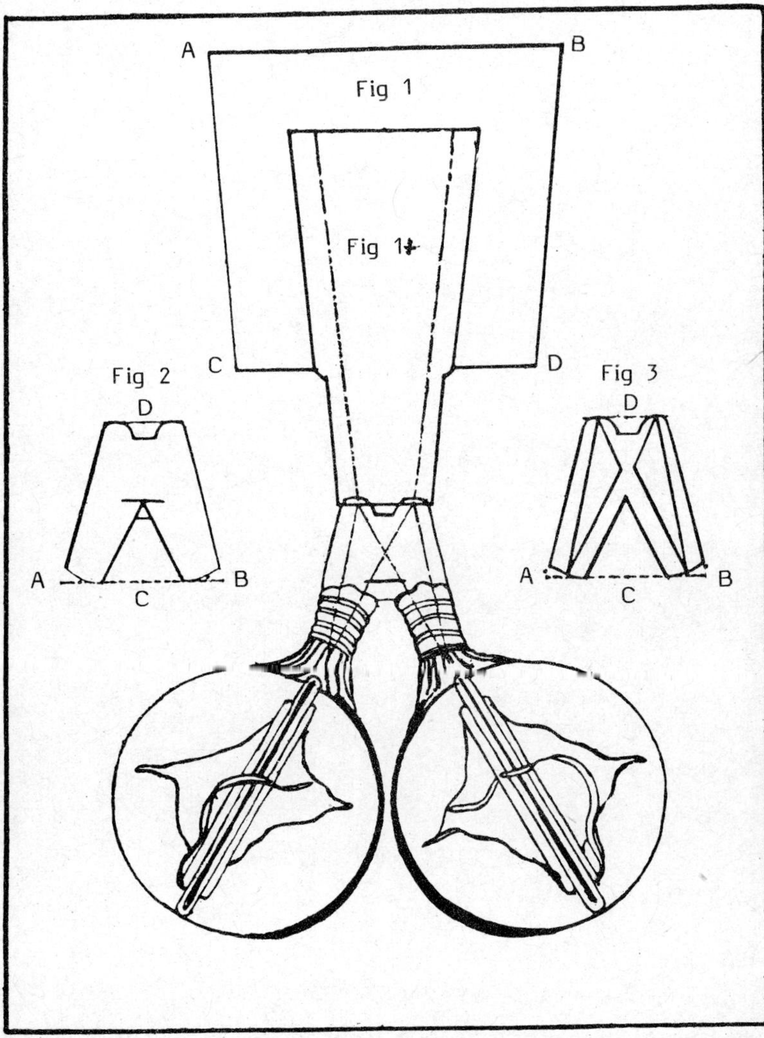

Geometrical Outline of the Furnace with Bellows.

PLATE 5

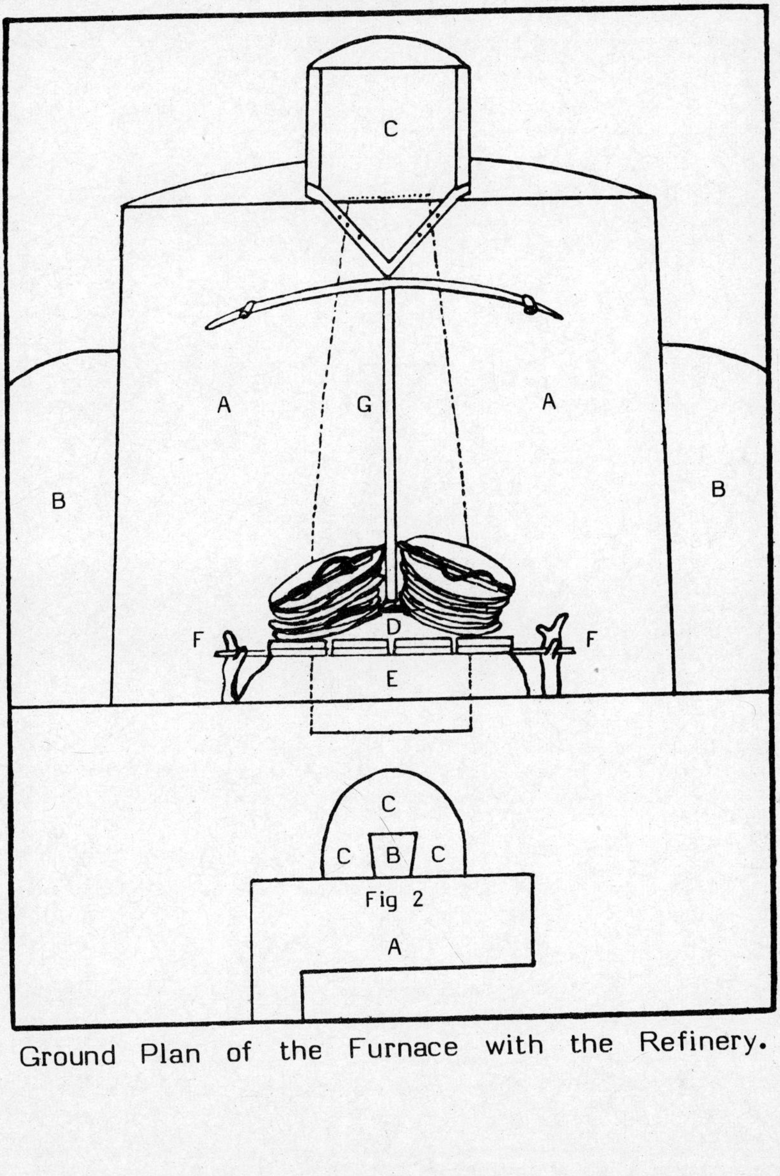

Ground Plan of the Furnace with the Refinery.

PLATE 6

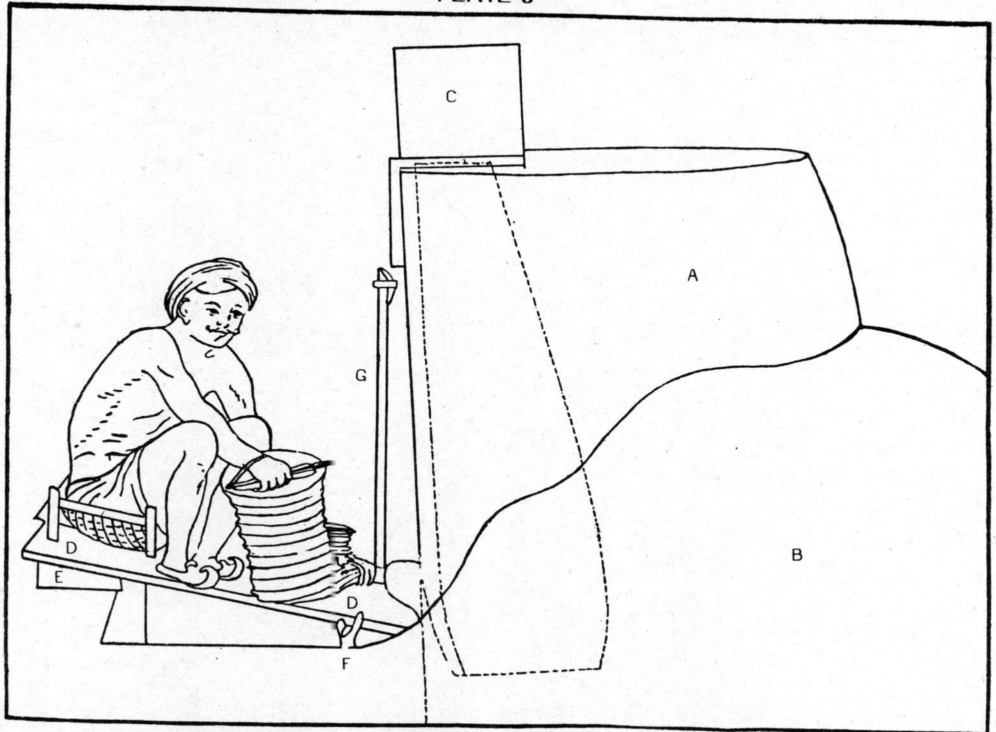

A Complete View of the Furnace with Bellows.

PLATE 7

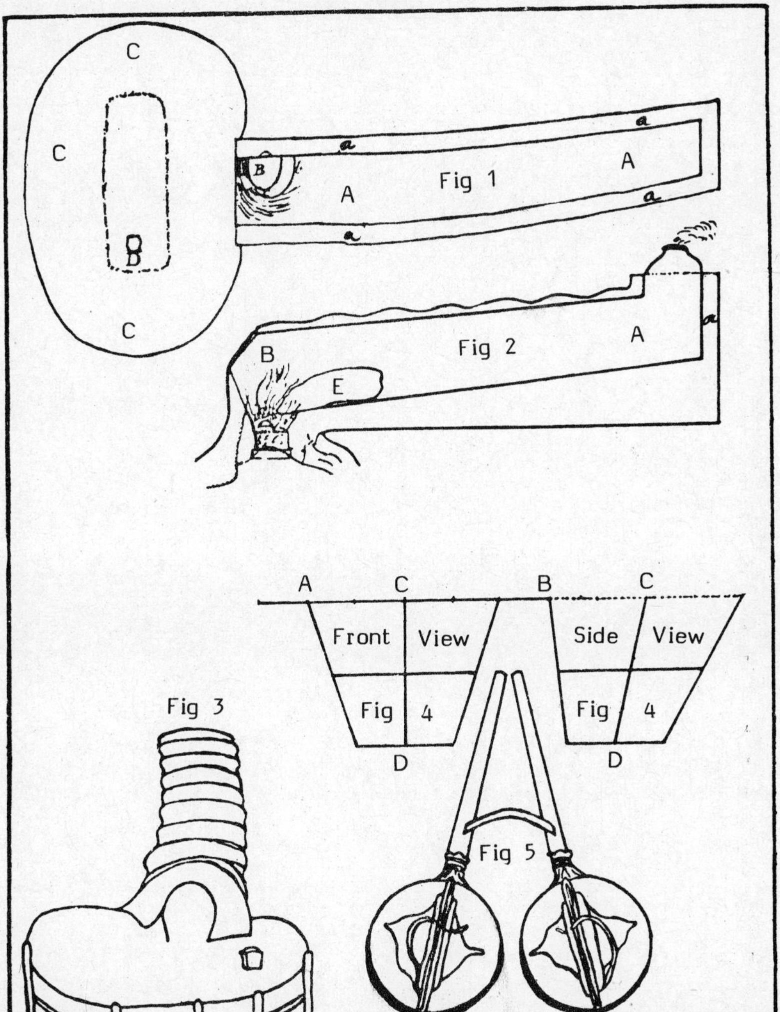

Geometrical Outline of the Furnace of the Refinery.

PLATE 8

A Complete View of the Refinery.

PLATE 9

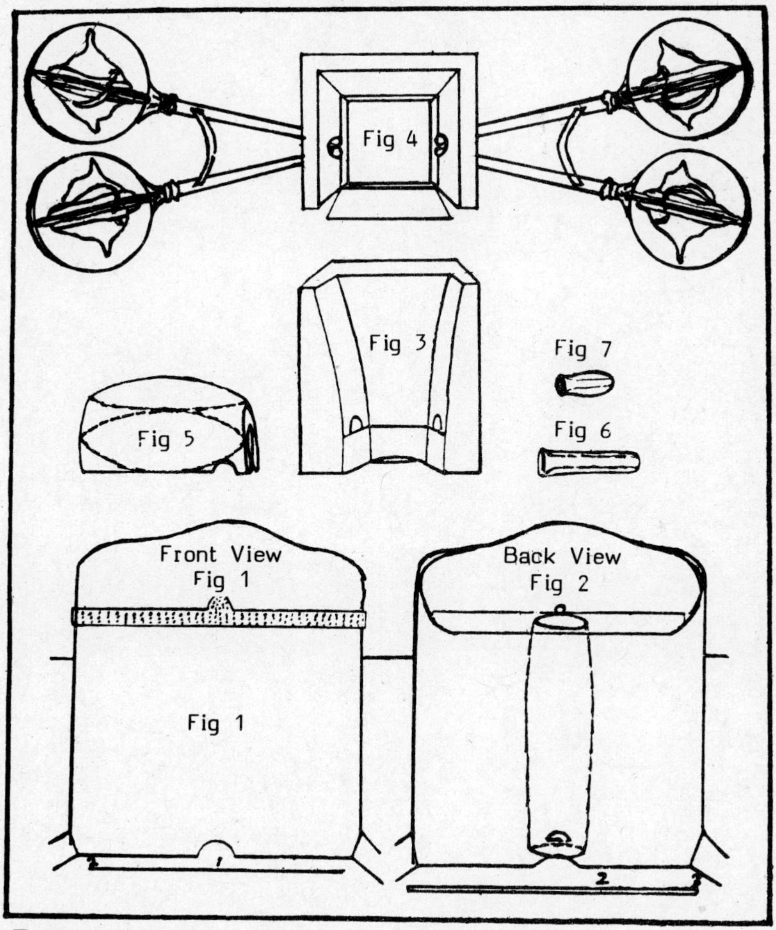

Front and Back View of a Small Circular Smelting Furnace.

PLATE 10

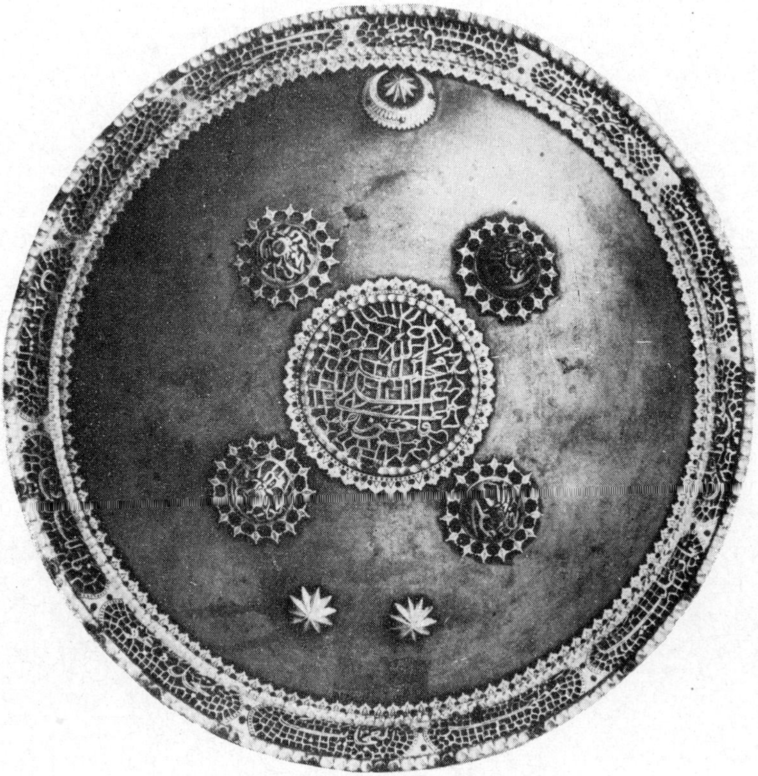

Steel Shield 18th Cent India.
(Courtesy, Salar Jung Museum, Hyderabad)

PLATE 11

Tega, Udaipur—Rajasthan, 18th Cent.
(Courtesy, Salar Jung Museum, Hyderabad)

PLATE 12

Jali Made of Cast Iron, Bijapur, Late 16th Century
139.7 X 86.4 cm.
(Courtesy, All Saints Church, Bijapur).

PLATE 13

Iron Smelting in the Santhal Parganas (From Iron & Steel Work in the Province of Bengal-Watson)

PLATE 14

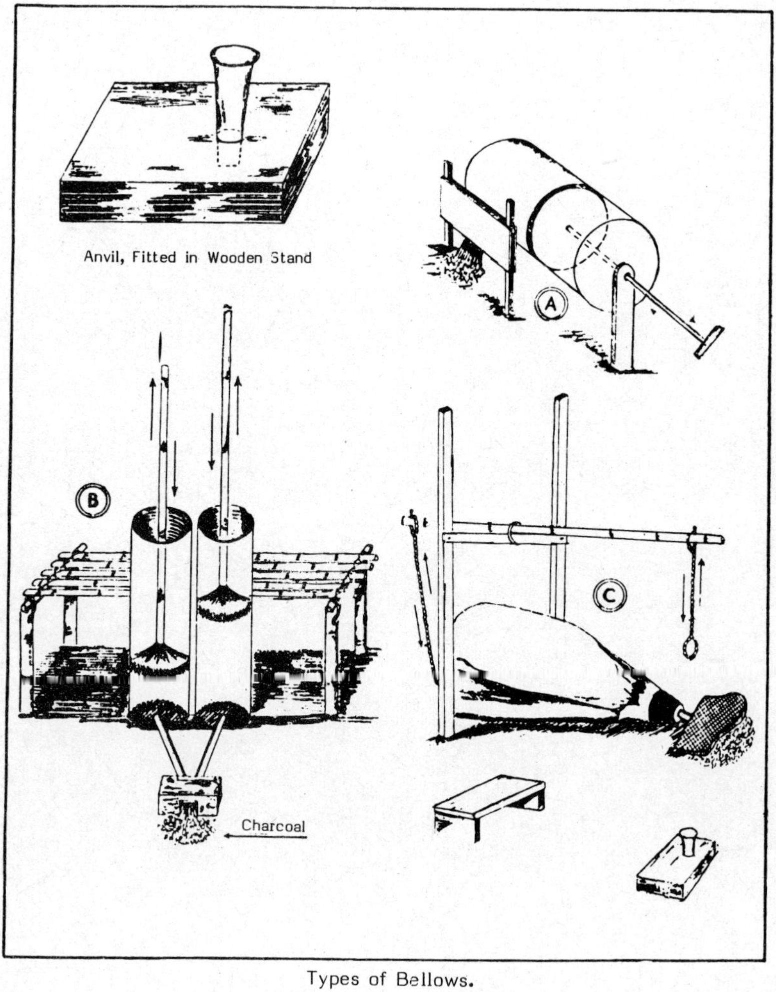

Types of Bellows.

PLATE 15

Dancing Girl, Mohenjodaro, Made By Cire Perdue Casting Method, 2500 B.C.
(Courtesy, National Museum, New Delhi)

PLATE 16

Nataraja Unfinished Statue, 12th Cent. A.D.
(Courtesy, National Museum, New Delhi)

PLATE 17

Natesa Chola Period 12th Cent. Made by Cire-Perdue Casting Method.
(Courtesy, National Museum, New Delhi)

PLATE 18

Inscription on the Fateh Rehbar Cannon at Golconda Fort.

PLATE 19

Silver Dagger Harappa 2700 B.C.
(Courtesy, National Museum, New Delhi)

PLATE 20

Akbar's Silver Coin, Agra Mint
(Courtesy, State Museum, Lucknow)

PLATE 21

Silver Vase 18th Cent.
(Courtesy, Salar Jung Museum, Hyderabad)

PLATE 22

Brass Copper Water Pot With Embossed Shikargah Scenes, Deccan 19th Cent.
(Courtesy, Salar Jung Museum, Hyderabad)

PLATE 23

Bidri Fakir's Bowl with Inscriptions, Deccan, India, Early 19th Century.
(Courtesy, Salar Jung Museum, Hyderabad)

PLATE 24

Bidri Huqqa Bottom 18th Cent.
(Courtesy, Salar Jung Museum, Hyderabad)